香港人氣甜品

Most Popular Desserts in Hong Kong

鄭慧芳 胡玉玲 編著

我寫過兩本書，第一本《幸福的甜品》主要介紹港式和台式的甜品，第二本《新手入廚學中菜》則以家常菜為主，感謝讀者的支持，今次再接再勵，揀選一些香港人氣甜品作骨幹，針對製作與處理部分，以讀者容易掌握為編寫宗旨。

香港人愛吃甜的東西，許多甜品轉瞬間便在香港大賣。香港是國際都會，除了中、西甜品外，泰式甜品也很受歡迎。提起人氣甜品，大家可能會聯想到潮流食品，但我卻認為人氣甜品不但要時尚合潮流，還要人們對它相當熟悉、有印象，以至每當被提及，隨即心領神會，想到在那裡吃的會較好。本書中的人氣甜品既包括特色的傳統食物，同時也有新興潮流的食品。

今次我與一位做糕點很認真的好朋友 Mindy 聯合編寫這本書。她努力地鑽研糕點，常在網上與人分享做餅心得和技術，深得朋友讚賞，更幫人做糕餅生意，因此她對做餅的工藝及材料的選用有着非一般的研究。而我在這幾年間，也不斷改良餅食及甜品的製作配方，終於能揣摩到一些簡單易明，適合家庭做的製作竅門，編集成書來跟讀者分享。

一如以往，我們希望能在書中詳細介紹一些取材方便、製作容易的食譜，並把一些食譜化繁為簡，希望連新手也可在家中做得出色。書中包含了隨時能為摯愛準備的冷熱甜品，也有為好朋友特做的糕點，務求讓你能從 "DIY" 中感受到幸福的滋味。

鄭慧芳

PREFACE 1

My first book *Chinese Desserts* is mainly about Chinese and Taiwan desserts and the second book, *Chinese Cooking for Beginners* is very popular for the easy and detailed cooking method. This book collects desserts in both traditional classic types and also new hit types in Hong Kong and the main theme of editing is focus on the making method which is easy for readers to handle.

In Hong Kong, this metropolis, no matter Chinese, western styles of dessert or even Thai one are very popular. Besides, Hong Kong people love desserts that make various kinds of desserts promptly hit. For me, mentioning about desserts will make me think about the related places to enjoy them.

This book is a cooperation between my friend, Mindy and I. Mindy is devoted in making confectionary and keen on improving the technique of making cakes. She often shares her experiences and skill online that makes her gets praises from her friends. Besides, she has her own confectionary business but keeps on exploring the art of making cakes and the selection of ingredients. About me, after years of improving skills, I finally found out some simple methods of making home-made confectionary to edit this book. As usual, Mindy and I want to dedicate simplified and easy-to-do receipes to all users so that even for beginners can handle it. Through the book, you can learn how to prepare different kinds of desserts for your loved one and feel the love and happiness.

Lilian Cheng

在2003年10月初，是我的烹調生活一個轉捩點。當年與摯友兼啟蒙老師鄭慧芳策劃做了一個自製鮮栗蓉配鮮忌廉的蛋糕，效果滿意，從此我陷入"無法自拔的狀況"。

自那天開始，家人眼中看到的我陷入瘋狂狀態，早上起床後要做蛋糕，到了深夜也不眠不休地做慕斯、曲奇、朱古力、奶凍、雪糕……製品中不是每個甜品都無師自通，有些會明查暗訪或慕名拜師。有時會為了蛋糕不夠香、不夠軟滑，徹夜不眠，就算失敗五次也好，十次也好，全都擊不倒我，甚至常發生在半夜2時起床"開檯做餅"的情形，如此不斷日復夜地實習，直至滿意為止。曾經因為我的狂熱和日夜顛倒的行徑，廚房一度被丈夫下了封鎖令，令我扼腕鼓噪，實是無可奈何……

正因有了長期練習和實踐的經驗，才發覺不停試煉那些來歷不明、道聽途說的食譜是虛耗光陰、浪費材料和精力的事，也意識到一本可靠而資料真實的學習手冊，十分重要，自此我對食譜挑選更嚴格了，漸而萌生正確記錄食譜的想法，這便是編寫這本《香港人氣甜品》的最初原因了。當然，可以得到好友鄭慧芳鼓勵，並與她一起鑽研改進，是一件愉快的事。書中每一款甜品都是精心挑選的，希望它能在你們的廚房裏佔有一席位。當然，書中難免有不足之處，希望大家不吝賜教！

胡玉玲

PREFACE 2

In early Oct of 2003, it was a turning point of my cooking life. Lilian, my mentor, also my buddy and I cooperated to make a DIY chestnut cream cake. The cake was so nice and then it drove me into a addictive situation.

During that period of time, I have been driven in a crazy condition, making different kinds of desserts like mousse, cookies, chocolate, pudding and ice cream etc...all day long. When I encountered difficulties, I will try best to find ways to solve them and sometime I desperately sought experts to teach me. I feared no failure and I will remake five times or ten times if the aroma, smoothness was not good enough. Waking up at the middle of the night for making cakes was almost routine pattern. Repeated practicing was my mission so as to achieve satisfaction. My freaky behaviour scared my husband who then stopped me stepping into the kitchen.....

After persistent practices, I understand that importance of a reliable, practical tool book and this is the motivation of editing this book "Timeless Popular Desserts in Hong Kong". With the help of my good friend, Wai Fong, every dessert in this book deploys our hearty effort and enthusiasm and is dedicated to all of you. Do not hesitate. Go make one now.

Mindy Wu

目錄 | CONTENTS

烤焗類
BAKED TYPES

煎、炸類
PAN-FRIED AND DEEP-FRIED TYPES

蒸、煮類
STEAMING AND BOILING TYPES

冷凍類
COLD TYPES

踏進 香港人氣甜品 的活動教室

STEPPING INTO THE WORKSHOP FOR POPULAR DESSERTS IN HONG KONG

焦糖脆脆天使蛋糕
Crispy Caramel **Angel Cake**

■ 製作時間：**120分鐘**
■ 模具：**8吋雪芳蛋糕模**
■ Production Time: **120 min**
■ Utensil: **8" chiffon cake mou**

⬊ Mindy 貼心小語

焦糖脆脆天使蛋糕，搭配簡單，看似平平無奇，卻是真正完美組合。鬆軟的蛋白雪芳蛋糕抹上香滑鮮忌廉，伴以焦糖脆脆同吃，口感豐富。那百搭的焦糖脆脆，用來配雪糕、奶凍或咖啡，同樣滋味無窮。有一股衝勁自己做吧！

This Crispy Caramel Angel Cake seems very simple but it is a perfectly matched one, soft and fine egg white chiffon cake coated by freshly made cream, with crispy caramel bites. Really have a drive to make one now, especially for the crispy caramel which can go with different food or drinks like ice cream, pudding or coffee.

材料
打起淡忌廉120克，打起甜忌廉120克

雪芳蛋糕
低筋麵粉90克（過篩），發粉1茶匙，大蛋黃3個
砂糖80克，菜油40克，鮮奶80克（加熱）
大蛋白6個，檸檬汁2茶匙或他他粉 1/4 茶匙

焦糖醬
砂糖100克，水2湯匙，熱水60克

焦糖脆脆
砂糖100克，麥芽糖25克
水25克，梳打粉5克

Ingredients
120g whipped cream (whisked)
120g sweet cream (whisked)

Chiffon Cake
90g soft flour (sifted), 1 tsp baking powder
3 egg yolk, 80g sugar, 40g vegetable oil
80g heated milk, 6 egg white
2 tsp lemon juice or 1/4 tartar powder

Caramel solution
100g sugar, 2 tbsp water, 60g hot water

Crispy caramel
100g sugar, 25g malt sugar
25g water, 5g baking soda

做法
雪芳蛋糕：

1. 粉類材料全部篩勻，加入砂糖40克、鮮奶、菜油及蛋黃一同拌至無粉粒，待用。
2. 把蛋白和檸檬汁（或他他粉）用攪拌器打起，分3次加入砂糖40克，續打至挺身，再把蛋白分3次徐徐捲入步驟（1）的粉糊中，輕輕攪拌至完全混合，倒入模具，置已預熱至170℃的焗爐中焗約35分鐘。出爐後倒扣出來，放涼。

焦糖醬：將砂糖加於水中，一邊用中火煮約3-4分鐘，一邊輕輕搖動煲具，直到焦糖呈現褐茶色，加入熱水，輕輕晃動煲子，離火放涼待用。

焦糖忌廉：把4茶匙焦糖醬拌入四分之一的淡忌廉內，置冰箱待用。

焦糖脆脆：砂糖、水及麥芽糖放小煲用中火煮約5分鐘，輕晃至混合，見焦糖呈現茶色便可離火，立即加入梳打粉，用木匙快手攪拌均勻，直至氣泡不斷膨脹，顏色變淺，便倒在已墊牛油紙的焗盤上，完全放涼後，輕輕壓碎待用。

組合：蛋糕橫切開半，在中間抹上焦糖忌廉，層疊起來，在表面抹上甜忌廉，置冰箱定型，吃時伴上焦糖脆脆。

Method
Chiffon cake:

1. Stir all flour, 40g sugar, milk, vegetable oil and egg yolk until combine.
2. Whisk egg white lemon juice (or tartar powder) by a mixing bowl, add in 40g sugar in 3 additions, stir until stiff peak. Fold egg white into (1) batter by 3 additions and mix. Pour into mould and bake at preheated oven at 170°C for 35 min. When baked, remove from mould and cool.

Caramel solution: add sugar into water and boil by medium heat for 3-4 min. Lightly shake the pot. When caramel turns to brown colour, add in hot water and lightly shake the pot. Remove from and cool.

Caramel cream: put 4 tsp of caramel solution into 1/4 whisked cream, chill in the fridge.

Crispy caramel: put sugar, water and malt sugar in a pot and boil by medium heat for 5 min. Lightly shake to combine. When solution turns to tea colour, remove from heat. Add in baking soda and stir to combine with a wooden spoon. When the caramel solution froths and turns lighter colour, pour on a baking tray lined with baking paper. Gently crush after cool.

Assembly: cut the cake into half horizontally, spread caramel cream in between to combine again. Spread sweet cream on top and chill in the fridge. Serve with crispy caramel.

美式芝士蛋糕
American **Cheese Cake**

↘ Mindy 貼心小語

美式芝士蛋糕是文華酒店的招牌貨。雖然品嚐過不同糕餅店的芝士蛋糕，但還是認為它最好，不花巧，只用簡單的材料，便能做到蛋糕香軟，芝士味濃質滑的效果，且毋須做餅底。

The American's Cheese Cake of Mandarin Hotel is very famous. It is subtle with simple ingredients but it can brings fine smooth texture, cheese flavour. After tasting different versions, I still perceive that the one of Mandarin is the best.

材料
牛油及麵粉適量（掃模具用）

忌廉芝士
芝士700克，室溫軟化
砂糖160克
無鹽牛油160克
雞蛋240克

蛋糕面
酸忌廉200克
砂糖40克

Ingredients
Butter and some flour
  (for brushing mould)

Cream Cheese
700g cheese (room temperature)
160g sugar
160g unsalted butter
240g egg

Topping
200g sour cream
40g sugar

■ 製作時間：**60分鐘**
■ 模具：**8吋圓形餅模**
■ Production Time: **60 min**
■ Utensil: **8" round mould**

做法

芝士蛋糕：

1. 在糕模內掃牛油，撲上麵粉。牛油隔水加熱，使之軟化。

2. 把芝士和砂糖攪打至滑身，分次加雞蛋拌勻，再加入牛油混合，倒進模具，在桌面上輕敲糕模，以釋出糕糊內的空氣。

3. 置預熱至160℃的焗爐中，焗40分鐘。取出，放涼。

蛋糕面：把酸忌廉和砂糖拌勻。

組合：把蛋糕面平整地抹在已放涼的蛋糕上，在冰箱放6小時便可進食。

Method

Cheese Cake:

1. Grease the mould with butter and sprinkle with some flour. Melt butter over hot water.

2. Whisk cheese and sugar until smooth, add in egg by several additions and mix to combine. Again add in butter and mix, pour into the mould. Lift up the mould and light hit the table so as to release the bubbles in batter.

3. Bake at the preheated oven at 160°C for 40 min. When baked, remove and cool.

Topping: mix sour cream and sugar.

Assembly: pour the topping over the cake and chill in the fridge for 6 hours.

Cook's Tips 技術指導

1. 這款芝士餅最好用手動打蛋器，如果用電動的，一定要用慢速，材料混合時便要立即停止，不要拂打太久，否則會把空氣一齊打進去，芝士餅便不夠細膩。

2. 爐溫過熱時，溫度相應調低一點。

3. 剛出爐的芝士蛋糕仍處未熟透狀態，餅中央仍未凝固，因此置室溫下，讓餘溫把蛋糕弄熟，恰到好處。

1. It is better to use hand whisk. If you use electrical one, turn to slow speed to mix up all ingredients. Do not whisk too long as air will go in which affects the softness of the cake.

2. When the temperature of the oven is overheated, lower the temperature.

3. After baked for 40 min, the cake still is not cooked completely and the core has not been solidified. Leave it in room temperature and let the heat of the cake continue to make the cake done.

朱古力香蕉蛋糕
Chocolate **Banana Cake**

■ 製作時間：**2.5小時**
■ 模具：**8吋圓形餅模**
■ Production Time: **2.5 hr**
■ Utensil: **8" round mould**

↘ Mindy 貼心小語

以簡約見稱，在人氣榜上排名坐亞望冠，看似貌不驚人，吃過的便知它內涵非比尋常。朱古力和焦糖香蕉絕佳組合，真是天造地設的食材。

This dessert is renowned for its simplicity and it hovers top one two on the favourite dessert list. Chocolate perfectly matches with caramel banana and you will know it is an unusual dessert after having a bite.

材料

朱古力蛋糕

朱古力80克，低筋麵粉110克，雞蛋3個
砂糖90克，牛油100克（室溫軟化）
梳打粉½茶匙，可可粉2湯匙，牛奶80克

香蕉餡

香蕉4隻（600克），檸檬汁30克，牛油100克
砂糖90克，魚膠片3克，冧酒10克

朱古力忌廉

打起忌廉220克，朱古力碎60克，鮮奶30克

朱古力淋漿

朱古力碎125克，淡忌廉150克
牛油15克，蜜糖2茶匙

Ingredients

Chocolate Cake

80g chocolate, 110g soft flour, 3 egg
90g sugar, 100g butter (room temperature)
½ tsp baking soda, 2 tbsp cocoa powder
80g milk

Banana Filling

4 banana, sliced (600g), 30g lemon juice
100g butter, 90g sugar
3g gelatine leaf (soaked in ice water), 10g Rum

Chocolate Cream

220g cream, whisked, 60g chopped chocolate
30g milk

Chocolate Sauce

125g chopped chocolate, 150g whipped cream
15g butter, 2 tsp honey

做法

朱古力蛋糕：

1. 香蕉切片；朱古力隔水加熱溶化；魚膠片用冷水泡軟；可可粉、低筋麵粉及梳打粉分別篩勻，備用。
2. 牛油及砂糖打至淡黃色，拌入可可粉，然後分數次加入雞蛋，拌勻，拌入半份粉材料，再拌入半份鮮奶，重複以上程序，最後拌入朱古力，置已預熱至170℃的焗爐，約焗40分鐘。

香蕉餡： 將牛油煮溶，加入已拌檸檬汁的香蕉，用中火煮約7-8分鐘，加入砂糖，待香蕉沾滿焦糖，加入冧酒，離火，加入魚膠片拌溶放涼備用。

朱古力忌廉： 鮮奶加熱，加入朱古力攪溶，放涼後拌入打起的淡忌廉，放冰箱備用。

朱古力淋漿： 淡忌廉加熱，加入朱古力及牛油拌溶，再放入蜜糖，放涼備用。

組合：

1. 把一片蛋糕塗上朱古力忌廉，放半份香蕉餡，再蓋上另一片蛋糕，重複以上程序，將整個蛋糕均勻抹上朱古力忌廉，置冰箱約30分鐘以定型。
2. 從冰箱取出蛋糕，放烤架上，將朱古力淋漿均勻地淋在蛋糕上，放回冰箱凝固定型，可隨意裝飾。

Method

Chocolate cake:

1. Melt chocolate over hot water. Sieve cocoa powder, soft flour and baking soda separately and set aside.
2. Cream butter and sugar until light yellow, add in cocoa powder and fold in egg by several additions and stir to combine. Fold in half portion of flour as well as ½ portion of milk and combine, again repeat this process and stir well. Finally add in chocolate, mix and pour into the mould and bake at preheated oven at 170°C for 40 min.

Banana filling: melt the butter and add in banana mixed with lemon juice, cook by medium heat for 7-8 min. Add in sugar and cook bananas until fully absorb caramel, pour rum and off the heat. Dissolve gelatine leaf in mixture and cool. Set aside.

Chocolate cream: boil milk and stir in chocolate, cool. Fold in whisked cream and chill in the fridge. Set aside.

Chocolate sauce: heat whipped cream and add in chocolate, melted butter and honey. Cool and set aside.

Assembly:

1. Spread chocolate cream upon a piece of cake and place half portion of banana filling. Cover it by another piece of cake. Repeat the above steps and spread chocolate cream on entire cake. Chill it in the fridge around 30 min.
2. Remove cake from fridge, place on wire rack and spread chocolate sauce evenly on it. Chill in the fridge again.

林明頓蛋糕
Lamington Cakes

材料

蛋糕

材料（A）
低筋麵粉55克，高筋麵粉55克
砂糖40克，發粉1茶匙

材料（B）
蛋黃（大）5個，鮮奶100克
沙律油50克，雲呢拿油少許

材料（C）
蛋白（大）5個，砂糖60克
檸檬汁2茶匙或他他粉¼茶匙

朱古力面

62%朱古力120克
淡忌廉120克，鮮奶60克
無鹽牛油20克，椰絲120克

Ingredients

Cake

Ingredients (A)
55g soft flour, 55g strong flour
40g sugar, 1 tsp baking powder

Ingredients (B)
5 large egg yolk, 100g milk
50g salad oil, some vanilla essence

Ingredients (C)
5 large egg white, 60g sugar
2 tsp lemon juice or ¼ tsp tartar powder

Chocolate Topping

120g chocolate (62% cocoa)
120g whipped cream
60g milk
20g unsalted butter
120g shredded coconut

■ 製作時間：**90分鐘**
■ 模具：**9吋正方形蛋糕模**
■ Production Time: **90 min**
■ Utensil: **9" square cake mould**

↘ Mindy 貼心小語

這一個是我極力推介的林明頓蛋糕，62%以上的法國品牌朱古力沾滿椰絲，一口咬下去，滿口香甜，加上雪芳蛋糕的鬆軟口感，肯定吃完一件再接一件，如果你嘗試過一些不甚滿意的林明頓，不妨試一試這個食譜，它一定會令你喜出望外。

I highly recommend this Lamington Cake which in 62% chocolate from France with shredded coconut and soft chiffon cake. You definitely cannot be satisfied by tasting one piece only. If you are not satisfied with those in coffee shop, you have to try to make one based on this receipe which can give you unexpected good result.

做法

蛋糕：

1. 先將材料（A）同過篩，然後把材料（B）的鮮奶加熱，再將材料（B）的其餘材料加入材料（A）中拌勻。

2. 把材料（C）的蛋白加檸檬汁或他他粉打起，分三次加入砂糖，續打至蛋白挺身，最後用打蛋器將步驟（1）的粉糊分三次混合拌勻，倒進糕模裏。

3. 焗爐預熱至170℃，放蛋糕糊焗35-40分鐘，出爐放涼，將蛋糕切成正方體待用。

朱古力面： 淡忌廉及鮮奶加熱至微滾便離火，倒入朱古力及牛油拌至完全融合，待涼。

組合： 將四方形蛋糕沾滿朱古力醬，黏上椰絲，放進冰箱定型，便成為一件件非常可口的林明頓蛋糕了。

Method

Cake:

1. Sift ingredients (A). Heat milk of ingredients (B) and then put all (B) ingredients into (A) and mix to even.

2. Whisk egg white of ingredients (C) with lemon juice or tartar powder, add in sugar by 3 additions and whisk until stiff peak. Again use a whisk to mix in (1) batter by 3 additins. Pour onto a mould.

3. Preheat the even to 170°C and bake the cake for 35-40 min. When baked, remove the cake from mould and cool. Cut the cake in cubes.

Chocolate topping: boil whipped cream and milk to very hot, pour into chocolate with butter and mix to even. Cool.

Assembly: stick the cakes with chocolate sauce and shredded coconut and chill in the fridge for setting.

心太軟
Warm **Chocolate Cakes**

■ 製作時間：**30分鐘**（不包括放冰箱時間）
■ 份量：**4杯**
Production Time: **30 min (not include chilling tim**
■ Yield: **4 pcs**

材料

雲呢拿雪糕4球，伴吃用
砂糖適量，焗杯用
牛油適量，焗杯用
糖粉適量，裝飾用

朱古力蛋糕

苦甜朱古力（70%可可成份較佳）
　　100克
無鹽牛油100克
冧酒1湯匙，砂糖50克
大雞蛋2隻，大蛋黃2隻
雲呢拿油½茶匙，低筋麵粉30克 ↖

Ingredients

4 scoops of vanilla ice cream
Some sugar (for baking cup)
Some butter (for baking cup)
Some icing sugar (for dusting)

Chocolate Cake

100g bitter sweet chocolate
　　(70% cocoa is better)
100g unsalted butter
1 tbsp Rum
50g sugar
2 large eggs
2 large egg yolk
½ tsp vanilla essence
30g soft flour

↘ Mindy 貼心小語

心太軟是朱古力迷的至尊甜品。我曾在某著名甜品店吃過變成朱古力蛋糕的心太軟，追問下才知點吃此甜品的客人實在太多了，朱古力漿來不及脫模，於是凝固了。但當時店方卻沒有抱歉或更換的意思，而我為了不掃朋友的雅興，只好當作是吃朱古力蛋糕。

Warm chocolate cake is the most favourite dessert for chocolate fans. Once I have ordered this dessert in a famous cake shop but the cake turned into a solid chocolate cake. It was found out that the baking time was over because of overwhelmed customers. In order to keep good appetites, we then enjoyed normal chocolate cakes!

做法

1. 焗杯掃上牛油，灑上砂糖。
2. 朱古力及牛油隔水加熱、溶化、拌勻，加入冧酒，備用。
3. 雞蛋、蛋黃、雲呢拿油及砂糖一同打成稠漿，加入朱古力溶液混合，然後篩入低筋麵粉，拌勻，倒入焗杯中約8分滿，放冰箱最少2小時。
4. 焗爐預熱至200℃，放入蛋糕焗約12分鐘，脫模後灑上糖粉，趁熱享用，伴上雲呢拿雪糕同吃，滋味更佳！

Method

1. Grease the baking cups with butter and sprinkle with sugar.
2. Melt chocolate and butter over hot water and stir to combine, add in rum and set aside.
3. Whisk egg, egg yolk, vanilla oil and sugar until thickened, add in chocolate mixture and mix. Sift soft flour onto the egg mixture and combine. Pour into baking cups about ⅘ full and chill in the fridge at least 2 hours.
4. Preheat the oven to 200°C and bake for 12 min, remove the mould and dust with icing sugar. Serve hot and with vanilla ice cream. It's fantastic.

Cook's Tips 技術指導

1. 朱古力漿能否流出來關鍵要視乎爐溫，因焗爐不同，溫度略有差異。
2. 蛋糕入爐數量會影響到烘焗時間，只要多做幾次，便能掌握自己的焗爐溫度，懂得調整時，一定會成功的。
3. 粉漿預先做好放冰箱，招呼到訪朋友，即時開爐烘焗，十分體面！

1. Be aware of the temperature control, it is the key of making the fluid chocolate filling.
2. The number of cakes baked affects the baking time, you can grasp the time control after several trails.
3. You can prepare the batter beforehead and keep it in the fridge. Then you can make this dessert whenever you like, especially treating your friend.

班尼士
Brownies

材料

低筋麵粉40克

可可粉10克

發粉1茶匙

朱古力200克，隔水坐溶

牛油100克，（室溫）軟化

大雞蛋2個

砂糖65克

鹽少許

冧酒／杏仁酒1湯匙

核桃100克，切碎

Ingredients

40g soft flour

10g cocoa powder

1 tsp baking powder

200g chocolate (melted over hot water)

100g butter (room temperature)

2 large eggs

65g sugar

some salt

1 tbsp Rum / Disaronno Amaretto (almond wine)

100g chopped walnut

■ 製作時間：**90分鐘**

■ 模具：**18厘米正方形模**

■ Production Time: **90 min**

■ Utensil: **18cm square mould**

↘ Mindy 貼心小語

某國際著名大酒店的招牌班尼士是以吋出售，有人説好吃，有人卻給予惡劣評價，道聽途説，不如自己親身體驗一下吧！只要買質量好一點的朱古力和可可粉做材料，照着我的方法去做，大家一定可以吃得盡情！

It is said that the famous brownies of an international hotel are sold by inches, some comment that it is great while some comment that it is bad. No matter it is good or not, it all subject to your experience. If you follow my instruction and purchase chocolate and cocoa powder in better quality, you surely reverse good rewards!

做法

1. 朱古力液及牛油拌至均勻。

2. 雞蛋、糖及鹽打至糖溶解（不用打至企身），加入朱古力牛油溶液及冧酒，攪拌均勻、表面出現光澤。

3. 粉類材料過篩，加入朱古力溶液拌至以不見乾粉粒為準。

4. 把朱古力麵糊倒入已墊牛油紙的模具中，刮平表面，灑上碎核桃。

5. 放在有水的焗盤裡，置已預熱至160℃的焗爐，焗約50分鐘，便可出爐。

Method

1. Mix melted chocolate and butter to combine.

2. Whisk egg, sugar and salt until sugar is dissolved (no need to achieve stiff peak), add in chocolate butter solution and Rum. Keep stirring until even and with shine.

3. Sift all flour onto chocolate solution and stir to combine until no flour shown.

4. Pour the chocolate batter onto the mould lined with baking paper. Level the top with sprinkle with chopped walnut.

5. Place in the preheated oven at 160°C and bake over a pan of water for 50 min. Remove when baked.

Cook's Tips 技術指導

1. 好吃的班尼士，烘焙秘訣是爐火控制，坐水盤焗50分鐘已經足夠，出爐後本身的餘溫會令蛋糕僅僅熟透，不會變得太乾。

2. 蛋糕涼凍後，放冰箱冷藏，凝固後切件，這樣不容易散碎，可以保持美觀。

3. 別忘記吃時稍放室溫更美味啊！

1. Secret of baking good brownies depend on the baking heat, it is sufficient to bake with a pan of water for 50 min. When baked, the remaining heat makes the cake just cooked and will not be too dry.

2. Cool the cakes and chill in the fridge until hardened, then cut and will not mash them.

3. Leave the cakes in room temperature for a while and serve.

杯子蛋糕
Cup Cakes

材料

杯子蛋糕
牛油（室溫）90克
砂糖 100克
鹽少許
打散雞蛋 2個
低筋麵粉 180克
發粉 1½ 茶匙
鮮奶 100克
雲呢拿油 ¼ 茶匙

原味牛油霜
糖霜 2杯，雲呢拿油 ¼ 茶匙
熱水 80克，牛油（室溫）50克
Crisco固體菜油 180克

Ingredients

Cup Cake
90g butter, room temperature
100g sugar
Some salt
2 egg, beaten
180g soft flour
1½ tsp baking powder
100g milk
¼ tsp vanilla essence

Original Butter Icing
2 cups icing sugar
¼ tsp vanilla essence
80g hot water
50g butter
180g Crisco shortening

■ 製作時間：**45分鐘**
■ 份量：**9個6cm紙杯**
■ Production Time: **45 min**
■ Yield: **6cm paper cup mould x 9**

↘ Mindy 貼心小語

杯子蛋糕熱潮方興未艾，只見不同款式和造型的杯子蛋糕陳列在咖啡室的櫥窗裏，真的可愛極了！心動不如手動，只要準備了杯子蛋糕、糖霜，再按喜好配上朱古力、咖啡、綠茶粉…或灑上可可粉、朱古力豆、棉花糖…發揮無窮創意，便能做獨一無二，玩味十足的蛋糕了！

Cup cakes prevail in the town. Eyed at different variety and design of cup cakes which make me feel excited as they are so cute! Get ready to make one now? Just need to prepare cup cakes and about icing sugar which depends on your favourite flavour like chocolate, coffee and green tea or even sprinkle cocoa powder or chocolate bits. All depends on your imagination and you can make your unique and fun cup cakes.

做法

杯子蛋糕：

1. 牛油、砂糖、鹽和雲呢拿油打成浮軟，分數次加入雞蛋，每次必需打至融合。

2. 加入 ⅓ 份鮮奶拌勻，然後加入已篩勻的麵粉，打勻，重複以上程序，直至所有材料完全融合為止。

3. 蛋糕糊倒入紙杯模約7分滿，置已預熱170℃的焗爐，焗約20分鐘，出爐後放涼。

原味牛油霜：

1. 糖霜用熱水調溶，加入雲呢拿油，待涼。如喜歡朱古力或咖啡味道，可添加1湯匙可可粉或咖啡粉便可。

2. 接着，加入牛油與固體菜油打至輕軟裝入唧袋，利用不同唧嘴，在杯子蛋糕面唧成上裝飾。

Method

Cup Cakes:

1. Cream butter, sugar, salt and vanilla essence until light and soft. Add in egg by several additions but make sure the mixture is completely combine before adding in next addition.

2. Stir ⅓ milk and add in sifted flour and mix to combine. Repeat this step until all has been stir to combine.

3. Pour batter into the paper cup mould around 70% full. Place in the pre-heated oven and bake for 20 min. Remove and cool.

Original Butter Icing:

1. Dissolve icing sugar in hot water and add in vanilla essence, and cool. If you prefer chocolate or coffee flavour, you can add 1 tbsp cocoa powder or coffee powder.

2. Then, add in butter and crisco and cream until light. Put into a piping bag and use different piping tubes to decorate different decoration.

焦糖咖啡芝士條
Caramel **Coffee Cheese Sticks**

材料

餅底
消化餅80克
核桃25克
牛油40克

焦糖芝士餅
忌廉芝士（室溫）500克
砂糖100克，特濃咖啡60克
雞蛋2個，蛋黃2個
焦糖醬20克，咖啡酒10克

Ingredients

Base
80g digestive biscuits
25g walnut
40g butter

Caramel Cheese Cake
500g cream cheese (room temperature)
100g sugar, 60g espresso coffee
2 eggs, 2 egg yolk
20g caramel sauce
10g Kahlua (coffee wine)

- 製作時間：**90分鐘**
- 模具：**7吋方形模具1個**
- Production Time: **90 min**
- Utensil: **7" square cake mould**

Mindy 貼心小語

我真的忘記了，這款芝士餅是從日本還是台灣傳來的呢？但它總是那麼受歡迎。把芝士蛋糕焗成方形，切成條狀，再用透明膠紙個別包裝好，用來送禮也很不錯呀！

I really forget the origin of this cheese sticks whether from Japan or Taiwan. But the fact is it is very popular in town. If you make this cheese cake in square shape, then cut into stripes and pack by individual clear bag with some nice decoration. It could be a good idea for making a gift!

做法

餅底：

1. 把消化餅放進保鮮袋裡，壓碎；核桃烘香、切碎；牛油隔熱水坐溶。
2. 在模具上掃牛油，灑上麵粉。
3. 把所有餅底材料拌勻放入模具中，用大木匙壓實，厚度要平均。
4. 置預熱至150℃焗爐焗8-10分鐘，取出備用。

芝士條：芝士加砂糖打至滑身，一邊拂打一邊分次序加入雞蛋、蛋黃、特濃咖啡、焦糖醬及咖啡酒，拌勻後倒入餅底。

組合：把芝士混合物置預熱160℃的焗爐，烘焗40-45分鐘，放涼，置冰箱冷凍6小時，切條。吃前，餅面可另淋上焦糖醬，味道更好。

Method

Base:

1. Crush biscuits in a plastic bag with rolling pin. Bake and crop walnuts. Melt butter over hot water.
2. Grease the mould and dust with flour.
3. Put all the base ingredients into the mould and use a wooden spoon to press firm but make sure achieving even thickness.
4. Bake at the pre-heated oven at 150°C for 8-10 min. Remove and set aside.

Cheese Sticks: cream cheese and sugar until smooth. While beating the cheese mixture, add in egg, egg yolk, espresso coffee and coffee wine in order and stir to combine. Pour it over the cake base.

Assembly: bake the caramel cheese mixture at preheated oven at 160°C for 40-45 min. When baked, cool then chill in the fridge for 6 hours. Cut into stripes and served with caramel sauce which enhances the taste.

Cook's Tips 技術指導

咖啡只是個人喜好，可以把咖啡酒轉成橙酒；焦糖漿和特濃咖啡則可轉為檸檬汁、藍莓醬、朱古力……隨自己的心情變化吧！

Is coffee your cup of tea? If not, you can change the flavour from coffee wine, Kahlua to Grand Marnier while turning syrup and espresso to lemon juice, blueberries or even chocolate…up to your mood.

英式楓糖鬆餅
Maple Syrup Scones

材料
低筋麵粉230克，發粉10克
砂糖25克，鹽少許
牛油75克
楓葉糖漿2湯匙（約36克）
淡忌廉120克
金提子乾（用熱水浸軟）80克

掃面
鮮奶少許
楓糖漿少量

Ingredients
230g soft flour
10g baking powder
25g sugar, some salt
75g butter
2 tbsp maple syrup (about 36g)
120g whipped cream
80g golden raisins, soaked to
　　soft in hot water

Garnishing
A little milk
A little maple syrup

■ 製作時間：**60分鐘**
■ 份量：**8件**
■ Production Time: **60 min**
■ Yield: **8 pcs**

◥ Mindy 貼心小語

這不是半島酒店的英式鬆餅，只是個低糖、低脂的食譜，但整個餅充滿着香噴噴的楓糖味，不用塗果醬已經十分可口！

This is not the British scones from the Peninsular Hotel but a low-sugar and low-fat one but entire scone is covered with maple syrup fragrance. It is very delicious even without jam.

做法
1. 把低筋麵粉、發粉、砂糖和鹽篩勻；把牛油從冰箱取出，切成1厘米×1厘米的小方粒，用手指把所有材料搓捏成麵包糠狀。
2. 將楓葉糖漿及忌廉拌勻，加入牛油麵粉粒混合搓揉，再加入金提子乾搓成粉團，置冰箱15-30分鐘。
3. 把麵糰取出，壓成約2-2.5厘米厚，切成大小一樣的三角形，掃上鮮奶。
4. 置已預熱焗爐，用爐溫230℃焗15-18分鐘，出爐後，趁熱在餅面再掃上楓糖漿，更添風味。

Method
1. Sift soft flour, baking powder, sugar and salt. Take butter from fridge and cut into 1cm x 1cm cubes and knead with flour mixture forming bread crumbs.
2. Mix maple syrup and cream together and add in butter mixture and knead together. Add in golden raisins and knead into a dough. Chill in the fridge for 15-30 min.
3. Take out the dough and press as 2-2.5 cm thick, then cut into consistent triangle shape and brush with milk.
4. Bake at the preheated oven at 230°C for 15-18 min. Remove and brush with maple syrup while hot. This can enhance the unique aroma.

Cook's Tips 技術指導

1. 把已搓好的麵糰放入保鮮袋中，放在砧板下壓好，厚薄均勻，置冰箱待用。
2. 傳統鬆餅用圓形模印出，若把剩下的粉糰搓勻再做鬆餅，口感會較硬，所以建議切三角形或方形，便不會浪費。

1. Pout the kneaded dough into a plastic bag and press under a cutting board until in even thickness. Chill in the fridge and set aside.
2. Traditional scones are cut by round mould. Scones baked from leftover dough will be more stiff. Therefore, triangular or rectangular scones will not make too much leftover.

意大利果仁條
Biscotti

材料

牛油（室溫）100克
砂糖 140 克
大雞蛋 2 個
低筋麵粉 400 克
發粉 1½ 茶匙
杏仁酒 2 湯匙
烘香腰果碎 100 克
烘香開心果碎 60 克

Ingredients

100g butter (room temperature)
140g sugar
2 large eggs
400g soft flour
1½ tsp baking powder
2 tbsp Disaronno Amaretto (almond wine)
100g cashew nuts (roasted, crushed)
60g pistachio nuts (roasted, crushed)

- 製作時間：**60 分鐘**
- 份量：**24 件**
- Production Time: **60 min**
- Yield: **24 pcs**

做法

1. 牛油和砂糖打軟，分數次加入雞蛋打起，再加入杏仁酒，拌勻。
2. 加入篩勻的低筋麵粉、發粉、鹽、腰果碎及開心果碎，待完全混合搓成麵糰。
3. 把麵糰平分2份，放入保鮮袋內，每份搓成23厘米×10厘米的長形麵糰，放冰箱冷藏1小時。
4. 取出麵糰置已預熱至180℃的焗爐烘焙20-25分鐘，等色澤呈金黃色取出，稍放涼便把大餅切成12件。把果仁條的切口向上，回放入爐焗15-20分鐘，至兩面金黃色出爐，放烤架待涼，便可享用。

Method

1. Cream butter and sugar until light. Add in eggs by several additions and beat, then add in almond wine and stir to combine.
2. Add in sifted soft flour, self-raising flour, salt, cashew nuts and pistachio nuts and knead into a dough.
3. Divide the dough into 2 portions and put into plastic bags. Each knead into a 23 x 10 cm rectangular dough and chill in the fridge for an hour.
4. Remove the dough from fridge and bake at the preheated oven at 180°C for 20-25 min until the colour turning to golden. Remove and cool. Cut into 12 stripes. Re-bake the sticks with cut end facing upward for 15-20 min until both ends turned golden. Cool at the wire rack and serve.

⊾ Mindy 貼心小語

在香港要吃Biscotti，除了部份五星級酒店外，就只有美式連鎖咖啡店了。其實Biscotti很容易做，它是我家長期提供的餅乾，無論伴鮮奶或咖啡，早餐或下午茶，都十分合適，只要把它焗乾一點，便可存放在室溫下，保存一個月都不成問題，快些動手吧！這次焗杏仁、下次焗腰果、再下一次焗夏威夷果仁……保證人人喜歡。

If you want to taste biscotti in Hong Kong, it will be available in five-star hotels or only in those American's chain café. Actually, biscotti is very easy to make and it is something always available at my home. The reason is it is a nice biscuit to be with milk, tea and coffee no matter it is for breakfast, tea or whenever you like. Besides, it can be kept in room temperature for one month if you bake it a bit dryer. The ingredients can be in various choices too like almond, cashew nuts or even macadamia nuts. There should be one fit you.

煎・茶類

撻・酥類

冷凍類

無花果金提燕麥條

Fig, Golden Raisins and Oat Sticks

材料

燕麥片300克
黃糖160克
牛油160克
肉桂粉¼茶匙
紅莓乾20克
乾無花果粒80克
金提子80克

Ingredients

300g rolled oat
160g brown sugar
160g butter
¼ tsp cinnamon powder
20g dry cranberries
80g chopped dry figs
80g golden raisins

■ 製作時間：**60分鐘**
■ 份量：**1磅12安士**
■ Production Time: **60 min**
■ Yield: **1 pound 12 oz**

↘ Mindy 貼心小語

在分秒必爭的清晨裏，趕校巴、港鐵、上早課或開會，幸好昨晚焗了燕麥條，取兩條作早餐吧！既方便又健康，不用到超級市場買！

Make oat sticks and simply take two pieces in hand for breakfast when you rush for buses, MTR, classes or meeting. Healthy and convenient and do not need to buy them in supermarkets.

做法

1. 金提子先用熱水浸軟，瀝乾待用。
2. 在模具墊上牛油紙或不黏布，掃上牛油待用。
3. 把牛油和黃糖用慢火煮溶，加入其餘材料拌勻，倒入模具中，用湯匙壓結實，厚度要平均。
4. 放進已預熱190℃的焗爐中，烘焗約25-30分鐘，當顏色轉金黃便要關爐，燕麥條仍待在爐中放涼，取出，切成條狀，便可享用。

Method

1. Soak golden raisins in hot water until soft. Drain well and set aside.
2. Line the cake mould with baking paper and brush with butter.
3. Melt butter and brown sugar by low heat, add in other ingredients and pour onto the mould. Press firm by a spoon with even thickness.
4. Bake at preheated oven at 190°C for 25-30 min. Turn off the heat when the biscuits turns to golden colour. Keep the biscuits in the oven and cool. Remove and cut into stripes. Serve.

煎·炸類

蒸·煮類

冷凍類

Cook's Tips 技術指導

買材料時，要選大片厚身的 "Rolled Oat"，因為它被滾壓過的，質地較結實，適宜做燕麥條。不要選即吃或快熟的燕麥，遇熱時它們便會變軟，做不到理想的效果了。

Purchase larger size "Rolled Oat" as it has undergone pressing and rolling process which in stiffer texture and fit for making oat sticks. Do not purchase instant oats or quick oats as they will turn tender and can not achieve the satisfied result.

燕麥提子果仁曲奇
Oat, Raisins and Macadamia Nut **Cookies**

材料

牛油100克
糖粉40克
黃糖60克
燕麥120克
椰絲50克
低筋麵粉130克
梳打粉¼茶匙
金黃提子乾70克
烘香的腰果碎或核桃碎70克
鮮奶50克
燕麥（沾面用）

Ingredients

100g butter
40g icing sugar
60g brown sugar
120g oat
50g shredded coconut
130g soft flour
¼ tsp baking soda
70g dry golden raisins
70g cashew nuts / walnut
  (roasted, crushed)
50g milk
Some oat, for topping

■ 製作時間：**40分鐘**
■ 份量：**1磅8安士**
■ Production Time: **40 min**
■ Yield: **1 pound 8 oz**

Mindy 貼心小語

這款曲奇本來自美國。話説一對母女到Dollars旅遊，在一間名為Neiman Marcus Mall的Café吃到這款曲奇，非常欣賞，付了250美元買下這個食譜，公諸同好。我以此作藍本，省去朱古力，減了糖份，加入大量燕麥及椰絲，吃起來柔韌可口，加上浸過的提子乾軟綿綿，非常可口，含有豐富纖維。

There is a story about this US cookie. It is said a mother and her daughter went to a café in the Neiman Marcus Mall of Dallars. They ate this cookie and enjoyed so much. Finally they used U$250 to buy the receipe and share all the public. I use the receipe as sample with eliminating chocolate, reducing sugar quantity and adding a great deal of oat and shredded coconut. With the addition of tender raisins, the texture then becomes chewy and rich in fiber. This cookie is really yummy and high fiber.

做法

1. 將焗爐預熱至190℃。提子乾洗淨，用熱水浸軟，瀝乾。

2. 將牛油、糖粉和黃糖打至滑身。

3. 加入椰絲、燕麥、低筋麵粉及梳打粉拌勻，再加入提子乾及腰果核桃碎混合。

4. 倒入鮮奶搓成麵糰，分成24等份，搓圓按扁，上下沾滿燕麥（份量以外）

5. 放進已預熱焗爐，焗約15分鐘成金黃色，出爐放烤架待涼，便成為健康可口的燕麥提子果仁曲奇了。

Method

1. Preheat the oven at 190°C. Rinse the raisin and soak in the hot water. Drain well.

2. Whisk butter, icing sugar and brown sugar to smooth.

3. Add in shredded coconut, oat, soft flour and baking soda and mix together, add in raisins and cashew nut or walnut, and mix.

4. Pour milk into the mixture and knead into dough. Divide it into 24 portions and press flat and coat with oat (beyond the ingredients list).

5. Pour the mixture onto a try and bake at preheated oven for 15 min or until golden yellow. Put on the wire rack for and cooling.

特鬆蛋黃曲奇
Crispy Egg Yolk **Cookies**

材料

人造牛油200克
糖粉50克
煮熟大蛋黃2個
低筋麵粉130克
粟粉70克
雲呢拿油¼茶匙
鹽少許

Ingredients

200g margarine
50g icing sugar
2 egg yolk (large, cooked)
130g soft flour
70g corn starch
¼ tsp vanilla essence
Some salt

■ 製作時間：**45分鐘**
■ 份量：**1磅4安士**
■ Production Time: **45 min**
■ Yield: **1 pound 4 oz**

做法

1. 牛油加糖粉打至浮軟。
2. 加入已壓碎的蛋黃及雲呢拿油打勻。
3. 篩入低筋麵粉、粟粉及鹽，以刮刀輕手拌勻，然後搓成麵糰。
4. 放進唧袋，唧出你喜愛的花紋，放入冰箱15分鐘。
5. 置已預熱至160℃的焗爐，烘焗約18-20分鐘，出爐放烤架待涼便可。

Method

1. Cream butter and sugar until light and fluffy.
2. Add in mashed egg yolk and vanilla essence and mix together.
3. Sift soft flour, corn starch and salt in the egg mixture gently, and knead into dough.
4. Put the dough into a piping bag and pipe into your favourite shape, chill in the fridge for 15 min.
5. Bake at the preheated oven at 160°C for 18-20 min. Remove and cool at wire rack.

↘ Mindy 貼心小語

近期香港熱賣的曲奇已不是數十年來，家喻戶曉的藍罐牛油曲奇了，而是熊仔罐裝曲奇，它具有無須咀嚼，入口即溶的口感，十分討人喜歡。不過，我喜歡加入熟蛋黃，這樣能給外表簡樸的曲奇添上蛋香味，還有獨特的鬆化口感，與濃濃的牛油香最匹配了。我最愛伴上一杯紅茶，便可享受一個懶洋洋的下午。

When it comes to talk about the recent good sale cookies in Hong Kong, it is no longer belong to the historic Danish Butter Cookies but teddy bear cookies. Teddy bear cookies is a lovely cookies as it will melt in your mouth and you do not need to chew it. In my recipe above, I add in cooked egg yolk as it brings special egg flavour to cookies. Besides, it is definitely enjoyable to sip a cup of tea with this soft cookie enriched with egg and butter flavour for the afternoon.

芝麻曲奇
Sesame **Cookies**

■ 製作時間：**30分鐘（不包括放冰箱時間）**
■ 份量：**12安士**
Production Time: **30 min (not include chilling tim**
Yield: **12 oz**

材料

牛油（室溫）110克
糖粉50克
鮮奶2茶匙
芝麻醬10克
杏仁粉30克
低筋麵粉150克
日本黑芝麻3湯匙
日本白芝麻3湯匙
砂糖（撒面用）適量

Ingredients

110g butter (room temperature)
50g icing sugar
2 tsp milk
10g sesame paste
30g almond powder
150g soft flour
3 tbsp Japan black sesame
3 tbsp Japan white sesame
Some sugar (for topping)

做法

1. 牛油加糖粉打至鬆軟，呈淡黃色，加入芝麻醬、鮮奶及杏仁粉打勻，然後加入低筋麵粉及黑、白芝麻拌勻，搓成麵糰。

2. 麵糰搓成2厘米×4厘米的長方體，用硬咭紙固定形狀，置冰箱2小時。

3. 取出麵糰，於表面撒上砂糖，切成8毫米厚片，排放在已墊牛油紙的焗盆上，請預留空間以便麵糰膨脹。

4. 置已預熱至170℃的焗爐，焗約15分鐘，放烤架待涼，放入密封容器存貯。

Method

1. Cream butter and icing sugar until light and turn to light yellow. Add in sesame paste, milk and almond powder and mix to even. Again stir in soft flour, black and white sesame and knead into a dough.

2. Knead the dough into a 2 x 4 cm rectangular shape and use cardboard to fix the shape. Chill in the fridge for 2 hours.

3. Take out the dough and sprinkle with some sugar and cut into 8mm thick. Put on the baking tray lined with baking paper. Leave some place for the dough to prove.

4. Bake at the preheated oven at 170°C for 15 min. When baked, remove and cool on the wire rack. Store in the vacuum container.

↘ Mindy 貼心小語

這款曲奇紅起來，到底是明星效應還是名廚效應？其實不提也知道它來自九龍城一家精緻餅店。在整盒雜錦曲奇中，我最愛芝麻乾果皮曲奇了，現選了這款來試做，由於家人不愛果皮味，所以棄用了，而改用芝麻醬及日本芝麻，焗出來的曲奇較香脆精緻。芝麻有潤膚美顏功效，味道香濃，口感鬆脆，一定要試試。

I do not know which celebrity makes this tin of cookies and only know this person owns a dainty cake shop in Kowloon City. Among different flavours of this shop's assorted cookies, I love sesame and tangerine peel cookie the most and that is the reason I choose this cookie for demonstration. Due to my family do not like the taste of tangerine peel, I then skip this ingredient and replace by sesame paste and Japanese sesame and the product made is relatively crunchy and neat. It is said that sesame can nourish skin and hair. You must try this amazing cookie then.

珍寶朱古力曲奇
Jumbo **Chocolate Cookie**

材料

牛油（室溫）70克
砂糖40克
黃糖40克
雞蛋1個
雲喱拿香油½茶匙
低筋麵粉160克
梳打粉1茶匙
入爐朱古力粒100克
烘香夏威夷果碎100克

Ingredients

70g butter (room temperature)
40g sugar
40g brown sugar
1 egg
½ tsp vanilla essence
160g soft flour
1 tsp baking soda
100g chocolate bites
100g macadamia nut (roasted, crushed)

- 製作時間：**30分鐘**
- 份量：**1個**
- Production Time: **30 min**
- Yield: **1 pc**

↘ Mindy 貼心小語

曾經在某大酒店見到一個個約9吋直徑大的曲奇，用不同顏色的朱古力寫上祝福生日的字句，很多大人和小朋友圍觀着，十分吸引。

I have seen a 9" wide mega cookie in a hotel, which has colourful chocolate made blessing word on top. Many passers-by were attracted by it and took a look at it.

做法

1. 把焗爐預熱至180℃。
2. 將牛油、糖粉和黃糖打至鬆軟、顏色變淡。
3. 分3次加入雞蛋及雲喱拿香油，每次均打至完全混合。
4. 拌入已篩勻的低筋麵粉、梳打粉、果仁及朱古力粒，直至完全混合，放在牛油紙上，壓平，或做成自己喜愛的形狀，焗約15-18分鐘，出爐放烤架待涼，便可裝飾。

Method

1. Preheat the oven to 180°C.
2. Cream butter, sugar and brown sugar to pale colour and light.
3. Add in egg and vanilla essence by 3 additions and mix to combine for each adding.
4. Fold in sifted soft flour, baking soda, crushed macadamia nut and chocolate bites and mix together, place on the baking paper and press flat. You may make your favourite pattern. Bake for 15-18 min. Cool on wire rack and decorate.

Cook's Tips 技術指導

1. 這個巨型曲奇，做法真的非常容易，還可利用不同蛋糕模具造型，只要在入爐前印出喜愛的形狀即可。
2. 焗完後用絲帶綁上大蝴蝶結作裝飾，作為一個簡便的生日禮物，精緻極了。

1. It is very easy to make this mega cookie and you can use different cake mould to make your favourite cookie pattern.
2. Wrap the cookie with a large ribbon for decoration and you can present this as kind of gift which is so neat and cute.

特濃吉士泡芙
Rich Custard Puff

材料

牛油（軟身）60克
鹽少許，清水120毫升
篩勻麵粉90克，雞蛋2個

吉士餡

蛋黃2個，砂糖80克
粟粉30克，清水2湯匙
牛奶240毫升，雲喱拿油¼茶匙
打起淡忌廉180毫升

Ingredients

60g butter, soft
Some salt, 120 ml water
90g sifted flour, 2 eggs

Custard Filling

2 egg yolk, 80g sugar
30g corn starach, 2 tbsp water
240 ml milk
¼ tsp vanilla essence
180 ml whisked whipped cream

- 製作時間：**60分鐘**
- 份量：**8個**
- Production Time: **60 min**
- Yield: **8 pcs**

Lilian 貼心小語

傳統泡芙的有大、細、長或圓形，各有不同形態和名稱，但我們認識的多是餅店中的朱古力或甜忌廉泡芙。近年，來自日本的泡芙專門店bread Papa's引入很多不同口味的大泡芙，給大家帶來不小驚喜，由於新鮮製造，香氣撲鼻，令人垂涎。

Shape of traditional puff can be different like large, small, long or round and will have different naming. The kinds we normally encounter in cake shops are chocolate or sweet cream puff. In recent years, there came a Japanese chain puff shop, named Bread Papa's which provide various kind of puff flavour. This interest the public especially for its freshly made quality.

做法

1. 牛油、鹽和清水同放小煲內煮至牛油軟化，大滾後隨即倒入麵粉快速攪勻，離火，用木羹繼續攪拌至麵糰離煲邊，倒入鐵鍋內待涼。

2. 雞蛋打散，分三次把蛋汁加入粉糰內，每次用打蛋器攪拌至滑身，拌成糊狀，用兩湯羹弄成小球，排放已掃油焗盆上，預留充足空間待其脹大。

3. 放入已預熱210℃的焗爐內，焗20分鐘至完全脹大，改用170℃再焗15分鐘至金黃色，取出，剝開半口，回爐焗5分鐘至乾脆，取出，待涼透。

4. 粟粉用水開勻，把蛋黃與砂糖攪拌至淡黃色，加入粟粉水拌勻。牛奶放煲內煮熱，注入蛋黃漿，拌勻，倒回煲內煮熟，取出，待涼。

5. 用打蛋器攪拌奶黃至軟滑，徐徐拌入雲喱拿油及忌廉，放入擠袋，把奶黃餡填滿泡芙中，雪凍後灑上糖霜裝飾，可再加鮮果伴食。

Method

1. Place butter, salt and water in a pot and cook, stir in flour vigorously when butter mixture brings to full rolling boil, make sure butter is melted. Remove from heat and use a wooden spoon to stir until it is clear. Put in a metal bowl and cool.

2. Whisk the egg and pour it into the dough by 3 additions. Make sure eggs are absolutely absorbed before adding next addition and beat to mix by a whisk each time. When the mixture becomes a thick batter, use a spoon to make balls and place on greased baking tray. Leave room for expanding.

3. Put into the preheated oven at 210°C and bake for 20 min until fully expanded. Reduce the heat to 170°C and bake for 15 min more until golden. Remove and cut one end. Re-bake for 5 min until dry. Remove and cool.

4. Dissolve corn starch in water. Cream egg yolk and sugar until light yellow and add in corn starch solution and mix together. Boil milk in a pot and add in egg yolk mixture and mix together. Pour into the pot and bring to boil. Off heat and let cool.

5. Use a whisk to beat the custard to smooth and fold in vanilla essence and cream. Then put into a piping bag and squeeze into puff. Dust with icing sugar for decoration and can be served with fresh fruits.

叮噹紅豆餅

Doraemon **Red Bean Cakes**

材料

克戟

麵粉 120 克
泡打粉 1 茶匙
梳打粉 ½ 茶匙
砂糖 1½ 湯匙
蜜糖 1 湯匙
牛奶 50 毫升
清水約 50 毫升
雞蛋 1 個
牛油溶液 2 湯匙

紅豆沙餡

日式紅豆沙餡 1 罐
清水 3 湯匙，粟粉 3 湯匙

Ingredients

Pancake

120g flour
1 tsp baking powder
½ tsp baking soda
1½ tbsp sugar
1 tbsp honey
50 ml milk
50 ml water
1 egg
2 tbsp melted butter

Red bean filling

1 Japanese tinned red bean purée
3 tbsp water
3 tbsp corn starch

- 製作時間：**40分鐘**
- 份量：**6件**
- Production Time: **40 min**
- Yield: **6 pcs**

Lilian 貼心小語

這餅來自一個動畫人物——叮噹，它最愛的小食。兩片煎餅夾著紅豆沙餡，在香港許多餅店都有出售，很受小朋友歡迎。

This hot cake is originated from the favourite snake of a cartoon figure - Doraemon. This kind of stuffed hot cakes is available in many cakes shops in Hong Kong and many kids love it.

做法

克戟：

1. 麵粉、泡打粉及梳打粉同篩在大碗中，中央開一穴，加入砂糖、蜜糖、牛奶、清水及蛋汁攪拌成粉糊，最後拌入牛油溶液，待10分鐘。
2. 慢火將易潔平底鑊燒熱，注入1大湯匙粉漿，慢火煎至表面開始呈現氣泡，反轉另一面再煎至熟，取出，再煎另一片煎餅。

紅豆沙餡：粟粉與清水調開，加入豆沙內拌勻，同放小煲內煮至濃稠，待涼，分成6份備用。

組合：用兩塊煎餅夾上豆沙餡，便成大受歡迎之叮噹豆沙餅。

Method

Hot cakes:

1. Sift flour, baking powder and baking soda into a bowl. Make a well in centre and add in sugar, honey, milk, water and egg. Stir all ingredients into batter and fold in melted butter. Leave it for 10 min.
2. Gently heat a pan and pour a big spoonful of batter on it. Gently fry until frothed. Turn over the hot cake and fry until cooked. Remove and fry another.

Red bean filling: dissolve corn starch in water and stir in red bean puree to even. Boil the mixture until thickened. Leave it cool and divide into six portions.

Assembly: stuff red bean filling between two hot cakes and here comes the popular Doraemon red bean cakes.

Cook's Tips 技術指導

1. 剩下的豆沙餅可入袋放雪霜內保存，進食前只要略略翻熱，也十分美味。
2. 節省時間，可用大平底鑊同煎兩片。
3. 當煎餅出現氣泡後便可離火片刻，約10秒後才反轉，回火續煎，煎餅比較鬆軟及色澤較吸引。

1. Chill the leftover red bean filling into the fridge. Re-heat before serving.
2. Pan fry two hot cake at the same time.
3. Switch off the heat when the cake is frothed. Leave it for 10 seconds and turn it over. Turn on the heat and continue to fry. This brings tender texture and attractive color.

鮮果忌廉班戟
Fresh Fruits and Cream **Pancakes**

材料

麵粉80克

吉士粉1湯匙

砂糖1½湯匙

雞蛋1個

鮮奶150毫升

清水150毫升

餡料

打起淡忌廉200毫升

篩勻糖霜2湯匙

去皮香蕉或芒果適量

焗香杏仁片2湯匙

朱古力醬適量

Ingredients

80g flour

1 tbsp custard powder

1½ tbsp sugar

1 egg

150 ml milk

150 ml water

Filling

200 ml whisked whipped cream

2 tbsp sifted icing sugar

Some banana or mango

2 tbsp baked almond slices

Some chocolate

■ 製作時間：**30分鐘**

■ 份量：**8件**

■ Production Time: **30 min**

■ Yield: **8 pcs**

↘ Lilian 貼心小語

有些人分不出班戟和克戟，其實分法簡單，厚塊是克戟，可塗牛油糖漿伴食；薄片是班戟。在酒店吃自助餐時，鮮果班戟是我首選的甜品，除了多款不同的鮮果，還有香脆果仁任君選擇，其實很多餅店也有出售芒果班戟，但多用植脂忌廉。如果自己做也很簡單，有時我會以雪糕代替鮮忌廉，味道不錯。

Some people cannot tell the difference of pan cake and hot cake. How to know it? Easy...the thick one is hot cake and can serve with butter and syrup while the thin one is pan cake. While enjoying buffet in hotels, fruit pan cake is my first choice of dessert and there are various kinds of fruits and crunchy nuts to serve with. For mango pan cake which is available in many cake shops but mostly stuffed with non-daily cream. Actually, it is easy to make it yourself. But I will use ice cream to replace fresh cream and taste is not bad.

做法

1. 雞蛋拂勻，加鮮奶及清水拌勻。

2. 麵粉和吉士粉一起篩入大碗中，中央撥開一穴，加入砂糖及半份奶水，用蛋拂拌成滑糊狀，再加入餘下奶水拌勻，用篩過濾。

3. 慢火燒熱平底鑊，用紙抹少許油，注入適量粉漿，薄薄上勻整個鑊，回火用中慢火煎至班戟離鑊邊，取出，待涼，用紙蓋好備用。

4. 淡忌廉拌入糖霜調味，入冰箱備用；班戟放大碟上，包入鮮果、忌廉，再用焗香杏仁片及朱古力醬伴吃，更添美味。

Method

1. Whisk egg and add in milk and water to even.

2. Sift flour and custard powder into a bowl. Make a well in centre and add in sugar and half portion of water. Use a whisk to whisk the mixture into batter. Stir in balance water until even and sift the batter.

3. Gently heat and grease a pan with a little oil. Add in suitable quantity batter and spread evenly on the pan. Turn to medium low heat to fry pan cake until cooked. Remove and allow to cool. Cover by paper and set it aside.

4. Fold icing sugar into whipping cream and chill it in the fridge. Place a piece of pan cake on a large plate and spread fresh fruits and cream on it. Roll up and serve with baked almond slices and chocolate paste.

牛油窩夫
Butter **Waffles**

材料	Ingredients
牛油（坐溶）50克	50g melted butter
酸忌廉80克	80g sour cream
雞蛋80克	80g egg
鮮奶200克	200g milk
低筋麵粉200克	200g soft flour
砂糖25克	25g sugar
酵母4克	4g yeast
鹽少許	Some salt
牛油（掃窩夫機用）適量	Some butter for brushing

■ 製作時間：**30分鐘**
■ 份量：**4塊**
■ Production Time: **30 min**
■ Yield: **4 pcs**

做法

1. 低筋麵粉、砂糖及鹽一同過篩，加入酵母備用。
2. 牛油、酸忌廉、雞蛋及鮮奶拌勻，分數次加入步驟（1）中，完全混合，用保鮮紙緊蓋粉漿約1小時。
3. 粉漿發酵完成時，表面會充滿氣泡，而體積比原身約大2倍。
4. 預熱窩夫機，掃上牛油，倒入粉漿，焗10-12分鐘便成香噴噴的牛油窩夫了。

Method

1. Sift the flour, sugar and salt and add in yeast. Set aside.
2. Mix butter, sour cream, egg and milk. Add in flour by a few additions and mix to even. Use cling film to cover it for an hour.
3. Batter is ready while frothing, size is triple as original.
4. Pre-heat waffle machine and brush with butter. Pour batter on it and bake for 10-12 min.

↘ Mindy 貼心小語

在香港，許多人都吃過窩夫，無論是塗滿牛油、花生醬、煉奶和砂糖的港式格仔餅，或是那曾經風摩一時的比利時牛油窩夫，都有大批捧場客。今次介紹這個是以比利時窩夫作藍本，減少了牛油的份量，加入了酸忌廉，吃起上來內鬆外脆，淋上蜜糖、朱古力醬、伴着雲喱拿雪糕、一口咬着士多啤梨和藍莓……噢！這頓下午茶就是它了。

In Hong Kong, many people have already tasted waffles no matter it is kind of local one with butter, peanut butter, condensed milk and sugar or that prevailed Belgium butter waffles and of course there are lots of waffle lovers. This time, let me introduce Belgium one with less butter but going with sour cream. Hmm…inner part is tender but outer part is crunchy, spreading with honey, chocolate paste, vanilla ice cream, fresh strawberries and blueberries………that's what high tea, go for a bite.

南瓜煎軟糍
Pan-fried Pumpkin **Rice Cakes**

材料
南瓜肉80克
砂糖1湯匙
生油1茶匙
粘米粉2湯匙
糯米粉90克

餡料
紅豆沙80克

Ingredients
80g pumpkin
1 tbsp sugar
1 tsp oil
2 tbsp rice flour
90g glutinous rice flour

Filling
80g red bean purée

■ 製作時間：**30分鐘**
■ 份量：**8件**
■ Production Time: **30 min**
■ Yield: **8 pcs**

Lilian 貼心小語

很多人都喜歡用糯米粉製成甜品，那軟糯口感，令人滿足。

Many people like to make dessert made up of glutinous rice powder, being contended by that soft and chewy texture.

做法
1. 南瓜去皮及籽，切厚片，放鐵碟內隔水蒸約10分鐘至軟腍，取出，隔去水及壓成蓉。

2. 南瓜蓉趁熱加糖、油、粘米粉及半份糯米粉先搓勻，然後搓入餘下糯米粉，揉成一不沾手粉糰，備用。

3. 紅豆沙搓長，分成8份，搓圓。

4. 粉糰搓長，分成8等份，每份包入豆沙，搓成圓形，再用手輕輕壓平。

5. 用易平底鑊燒熱3湯匙油，下南瓜軟糍蓋好用中慢火先煎1分鐘，反轉再煎1分鐘，然後開蓋繼續多煎1-2分鐘至兩面金黃色，取出，用吸油紙吸去多餘油份，排放碟上趁熱享用。

Method
1. Skin and remove seeds of pumpkin and cut into thick slices. Place it on a metal plate for steaming around 10 min until softened. Remove, drain well and mash into purée.

2. Add in sugar, oil, rice flour and half portion of glutinous rice flour whilst pumpkin is hot. Mix all ingredients to combine. Fold in balance glutinous flour and knead until forming a clear dough. Set it aside.

3. Roll the red bean purée and cut into 8 portions. Knead into balls.

4. Roll the dough and cut into 8 portions. Stuff one red bean purée into each and knead into balls. Gently press to flat.

5. Heat the pan with 3 tbsp oil and pan-fry rice cake with lid cover. Pan-fry the rice cake with lid covered for 1 min by medium heat. Turn over the cake and pan-fry 1 min too. Remove the lid and continue to pan-fry for 1-2 min until golden. When cooked, remove and absorb excess oil. Dish up and serve whilst hot.

Cook's Tips 技術指導

當煎南瓜軟糍時，注意油溫不可太高，煎時要蓋住軟糍，以便焗熟，及後，開蓋再煎，否則表面不夠香脆，會變得很油膩。

When pan-frying the rice cakes, pay attention to the temperature. It cannot be too high. Remember to cover by a lid to ensure the cakes are heated. Then remove the lid to pan-fry continuously, otherwise, the surface will not turn crispy but oily.

香煎百寶糯米飯

Pan-fried Glutinous Rice with **Eight Treasures**

材料

糯米300克（8兩）

紅豆沙150克（4兩）

松子仁2湯匙

杏脯肉2粒

提子乾2湯匙

元肉2湯匙

糖桂花1湯匙

豬油1茶匙

糖桂花適量

Ingredients

300g glutinous rice

150g red bean purée

2 tbsp pine nut

2 dried apricot

2 tbsp raisins

2 tbsp dried longan pulp

1 tbsp osmanthus sugar

1 tsp lard

Some osmanthus sugar

■ 製作時間：**50分鐘**
■ 份量：**6人**
■ Production Time: **50 min**
■ Serving: **6**

Lilian 貼心小語

百寶糯米飯是上海人食團年飯必備食品，以示一家團圓。其實八寶糯米飯在上海辦館很易買得到，回家蒸熱後便可食用。所謂"八寶"，一般包含有蓮子、蜜棗、松子仁、提子乾、還有紅紅綠綠的蜜餞車厘子等配料。若自己做，可選擇自己喜愛配料，還新鮮得多，是冬天最佳甜品，不妨一試。

Glutinous rice with eight treasures is an indispensable food for the meal of Shanghai's family reunion, representing reunion of family. This kind of dessert can be found in Shanghai grocery and is easy to treat simply by steaming. "Eight treasures" usually consist of lotus seed, raisins and those colourful sweetened cherry ingredients. If you want to make it yourself, you can choose your favourite fresh in gradients. It is kind of great dessert in winter time to share with your family.

做法

1. 糯米洗淨，用水浸約4小時至漲軟。

2. 紅豆沙壓平；杏脯肉切碎，與提子乾及元肉用熱水沖淨，備用。

3. 糯米隔乾水後放淺窩內，以大火蒸30分鐘，取出，趁熱拌入豬油、桂花糖、松子仁及提子乾等材料。

4. 將糯米飯分成兩份，預備保鮮紙鋪淺窩內，放一份糯米飯，再放上紅豆沙，用另一份糯米飯鋪面，用保鮮紙包裹好，壓成餅形，備用。

5. 八寶糯米飯可蒸熱食，或用平底鑊加油煎至兩面金黃香脆，切件上碟，再淋少許糖桂花伴食，更添美味。

Method

1. Rinse the glutinous rice and soak in the water for 4 hours.

2. Press the red bean purée, chop the apricots and rinse all together with raisins by hot water. Set aside.

3. Drain glutinous rice and put it in a shallow bowl and steam by high heat by 30 min. Remove and fold in lard, osmanthus sugar and mixed fruit whilst hot.

4. Divide the glutinous rice into two portions. Place one portion over a shallow pot covered with glide cline and put red bean purée on top. Again place balance glutinous rice on top and cover by glide cline. Press to form cake shape. Set aside.

5. Treasures Rice can be steamed or pan-fried until both sides turning golden. Chop the cake into pieces and serve with osmanthus sugar which enhances the flavour.

冰花蛋球
Chinese Styled **Donuts**

材料
麵粉 100 克
鹽 ⅛ 茶匙
清水 120 毫升
豬油或固體菜油 15 克
食用臭粉 ⅛ 茶匙
雞蛋 3 個

飾面
幼砂糖適量

Ingredients
100g flour
⅛ tsp salt
120 ml water
15g lard or Crisco shortening
⅛ tsp ammonia powder
3 eggs

Garnishing
Some sugar

■ 製作時間：**20分鐘**
■ 份量：**12件**
■ Production Time: **20 min**
■ Yield: **12 pcs**

Lilian 貼心小語

有人誤稱這甜點為沙翁，其實沙翁是沒有蛋香味，這個應該名為冰花蛋球，因這個炸球含豐富雞蛋成份，所以可帶出香濃蛋味及鬆軟口感。製作時要令它發至如此脹大，除了多蛋汁外，還要有臭粉的作用，但用臭粉份量要恰到好處，否則便有怪味。

Some people misunderstand this dessert is egg buns, however, it 5 brings no egg smell. The one that introduced now should be named as Chinese Styled Donuts as it has rich egg smell and tender texture. How to make the buns expand rely on egg fluid and ammonia powder. But pay attention to the quality used, if excess, it brings strange flavour.

做法
1. 麵粉與鹽同篩，備用。
2. 豬油及清水放煲內以慢火煮溶，開火煮至大滾，隨即加入食用臭粉及麵粉快速攪拌，離火，繼續攪拌成一熟粉糰，倒入鐵窩內備用。
3. 雞蛋拂勻，分5次將蛋汁加入粉糰內，可用打蛋器以慢速攪拌，直至攪拌成滑身麵漿。
4. 燒半鍋熱油，用兩隻沾過油的鐵羹把麵漿捏成荔枝般大小球形，放下熱油中，以中慢火炸至脹大及金黃色，期間蛋球會自動轉動，膨脹，直至蛋球將近分裂時即可撈出。
5. 取出後要即時滾上砂糖，上碟享用。

Method
1. Sift flour and salt. Set aside.
2. Boil lard and water in a pot by low heat until dissolved. Turn to high heat and stir in ammonia powder and flour until combine. Remove from heat and stir to form dough. Move to a metal pot and set aside.
3. Whisk eggs and add in the dough by 5 additions. Stir the mixture by low speed until forming batter.
4. Heat half pot of oil. Use two greased metal spoons to make balls in lychee size and put the balls in the pot. Deep fry the egg balls by medium heat and balls will self-roll, expand and turn to golden colour. Remove while almost split.
5. Roll on the sugar and serve.

Cook's Tips 技術指導

臭粉多為點心專用，也可在烘焙店舖買到，如果找不到也可用泡打粉代替，效果也算不錯。

Ammonia powder is usually for making dim-sum and can be found in bakery shops. If you cannot find this powder, you can replace it by baking powder the effect is not bad.

拔絲香蕉
Banana **Fritters**

材料
半熟香蕉2隻
麵粉2湯匙
炒香黑芝麻1茶匙
冰粒適量，冰水適量

粉漿
麵粉90克
生粉1湯匙
梳打粉¼茶匙
蛋白2湯匙
冰水60毫升

糖膠
生油1湯匙
砂糖150克
清水60毫升

Ingredients
2 skinned, semi-ripe banana
2 tbsp flour
1 tsp roasted black sesame
Some ice cubes, some ice water

Batter
90g flour
1 tbsp corn starch
¼ tsp baking soda
2 tbsp egg white
60 ml ice water

Syrup
1 tbsp oil
150g sugar
60 ml water

■ 製作時間：**20分鐘**
■ 份量：**4人**
■ Production Time: **20 min**
■ Serving: **4**

☒ Lilian 貼心小語

拔絲蘋果、拔絲香蕉是北方
特色甜點，在香港的北京樓
一定可以吃得到。其實用番
薯或芋頭來做拔絲菜味道也
不錯，因不會太甜。

Apple fritters and banana
fritters is a unique northern
dessert. You can taste it
in a local restaurant called
Peking Garden Restaurant.
Actually, you can replace
the fruit filling by potato or
yam and the taste is pretty
good and not too sweet.

做法
1. 把三種粉材料同篩大碗中，徐徐拌入蛋白及冰水至幼滑粉漿，備用。
2. 香蕉去皮，切厚件，即時灑上麵粉，再沾上粉漿，放熱油中炸至淺黃色，取出，隔油。
3. 用小煲將糖膠材料煮滾，期間不要攪動，以中慢火煮5-7分鐘，直至糖膠呈金黃色，可用筷子沾一滴糖膠滴進水中，若糖膠立刻變硬便可把已炸好的香蕉續粒沾上糖膠及浸入冰水，待糖膠變硬，放回已塗過油碟上。
4. 灑上芝麻裝飾，趁熱享用。

Method
1. Sift three kinds of flour in a large bowl. Lightly fold in egg white and ice water, mix to smooth batter. Set aside.
2. Cut banana into thick pieces, sprinkle with flour and dip with batter. Deep fry until light golden. Remove and drain.
3. Use a small pot to boil syrup ingredients and do not stir by medium heat for 5-7 min until turning to golden colour. Use a chopstick to dip a drop of syrup into water. If it turns to hardened, dip all fried banana with syrup and put into ice water. Once syrup is hardened, remove to a greased plate.
4. Sprinkle sesame on top for decoration and serve hot.

Cook's Tips 技術指導

1. 做拔絲菜最重要是煮糖膠的溫度及時間，煮不夠火喉，糖漿太稀就拔不出絲，煮得太過火糖就變焦，吃起來有苦味，所以用水試糖膠也是個好方法。
2. 盡快把製品上糖漿後浸冰水，否則香蕉會因冷卻纏在一起而分不開。

1. The key point of making this dessert is the temperature and time of making syrup. If processing time is too short, syrup is too thin and cannot make stretchy syrup effect. But if the time is too long, the syrup is overcooked with bitter taste. It is a good method to test syrup effect by water.
2. Speed up the dipping process in syrup and followed with ice water, otherwise, banana will cool and stick together.

反沙芋條
Candied **Yam**

<table>
<tr><td>

材料
大芋頭半個
砂糖 120 克
清水 50 毫升

</td><td>

Ingredients
Half yam
120g sugar
50 ml water

</td></tr>
</table>

■ 製作時間：**20分鐘**
■ 份量：**4-6人**
■ Production Time: **20 min**
■ Serving: **4-6**

做法

1. 芋頭去皮沖淨，修切出12條約6厘米×2厘米的厚條。
2. 用煲燒2杯熱油，放下芋條炸6-8分鐘，直至表面皮硬及呈微金黃色，取出。
3. 同時，用另一鑊燒1茶匙油，加水及糖以慢火煮6-8分鐘，煮時可攪拌糖膠，直至糖膠呈濃泡狀，可取一滴糖膠滴進凍水中測試硬度，若變硬便可。
4. 芋頭條加入糖膠內隨即熄火，快手兜炒3-4分鐘直至外面呈現白色砂糖狀，上碟，趁熱品嚐。

Method

1. Skin the yam and rinse. Cut it into 12 pieces (6 x 2 cm thick).
2. Heat 2 cups of oil in a pot, deep-fry yam for 6-8 min until skin is hardened and light gold. Take out.
3. Heat 1 tsp of oil in another pot, add in water, sugar and boil by low heat for 6-8 min. Stir occasionally until extreme frothed. Dip a drop of syrup into cold water for hardness testing. If it is hard, done.
4. Put yam into syrup and turn off the heat. Stir fry yam quickly for 3-4 min until coated by white sugar grains. Dish up and serve hot.

◥ Lilian 貼心小語

反沙芋條是潮汕名點，九龍城的一間傳統潮州菜館做得很美味，但價錢相當昂貴，提議大家在家中製作。

This candied yam is a renowned Chiu Chow's dessert. A traditional Chiu Chow's restaurant in Kowloon City is good at making this dessert but very expensive. With the help of this receipe, all of us can try to make this delicious dessert by yourself.

Cook's Tips 技術指導

1. 首選大個靚芋頭，泰國芋頭品質較有保証，最好用芋頭中心部位最為鬆化軟腍。
2. 反沙的奇妙處，正當糖水煮成糖膠狀態，如果沒有攪動過，冷卻後便成硬糖。相反地，經過不停兜炒，因加入了空氣，冷卻後的糖膠會還原成砂糖，沾在芋條上的糖是鬆脆的。

1. Choose large yam especially that is Thailand origin with better quality. Choose the middle part as it is the most softened part.
2. It is amazing that no stirring of sugar makes hardened syrup, conversely, stirring allow air to go in which turns syrup back to sugar grains form which is crispy.

蛋黃蓮蓉大壽包
Peach Bun with **Lotus Seed Paste**

材料

皮料
麵粉200克，泡打粉2¼茶匙
蛋白1個，幼砂糖50克
豬油3茶匙，鮮奶40毫升
清水約40毫升

餡料
白蓮蓉240克
鹹蛋黃6個

飾面
桃紅色素少許食用
清水少許

Ingredients

Dough
200g flour
2¼ tsp baking powder
1 egg white, 50g sugar
3 tsp lard, 40 ml milk
40 ml water

Filling
240g white lotus seed paste
6 salted duck egg yolks

Decoration
Some red food colouring
Some water

■ 製作時間：**90分鐘**
■ 份量：**6個**
■ Production Time: **90 min**
■ Yield: **6 pcs**

↘ Lilian 貼心小語

大家在壽宴酒席多會吃到壽包，自從東海酒家用原個蛋黃大壽包作招徠後，大家開始對壽包另眼相看，更有酒樓出品大壽包內藏小壽包，非常可愛。不過，最有誠意還是自己親手做，我這個是簡易製大壽包，有愛心人士一定做得到。

Usually, you can taste this peach bun in birthday banquet. However, it seems there is a change after launching of mega-peach bun with entire salted duck egg yolk by East Ocean Seafood Restaurant. People are attracted by different modified cute designs like a mega-peach bun contains small buns inside. Truly, it will be more precious if the bun is made by myself. so, just follow the easy-to-make recipe and make one yourself.

做法

1. 麵粉及泡打粉同篩大碗中，中央撥開一穴，加入蛋白、砂糖、豬油、牛奶及清水，徐徐從中央拌成濃糊狀，再把乾麵粉拌入，搓成粉糰，用布蓋好待40分鐘。

2. 鹹蛋黃沖水後抹乾；白蓮蓉分切成6份，每份包入一粒鹹蛋黃，搓圓。

3. 麵粉糰分切成6份，每份粉糰灑上少許麵粉，用木棍輾成12厘米×5厘米長條，向上捲成圓筒形，再合上成方塊，壓平，再輾成6厘米圓塊，皮邊要輾得較薄，包入蓮蓉餡，用手將麵皮向上捏，收口，多餘麵皮要去除，成包底。

4. 在包頂處再用手向上捏成挑尖形，放包紙上，置蒸籠中以大火先蒸5分鐘，逐個用尖頭筷子從底部向上壓，成桃包紋，回火繼續蒸15分鐘，期間每5分鐘開蓋一次，以防包皮因過熱而爆裂。

5. 用牙刷蘸少許色素液，用刀撥動牙刷，彈向包皮作裝飾，便成大壽包。

Method

1. Sift flour and baking powder in a bowl. Make a well in centre and add in egg white, sugar, lard, milk and water. Lightly fold all ingredients into batter. Add in flour and knead into a dough. Cover with cloth for 40 min.

2. Rinse salted duck egg yolks and rinse well, divide the lotus paste into 6 portions. Each stuff with a piece of egg yolk, then knead to a ball.

3. Divide the dough into 6 portions and sprinkle with some flour. Use a rolling pin to knead into a 12 x 5 cm paste, roll it up to form cylinder, then fold up to form a square and press flat. Again knead into 6 cm squares. The rim needs to be thinner. Stuff with lotus seed paste and knead to seal. Trim excess dough to form bottom part.

4. Knead the top of dough upward to form pointed peach shape and place the bun on bun paper. Put all buns on steamer and steam by high heat for 5 min. Then, use chopstick tip to press the dough from bottom to top, forming peach shape. Continue to steam for 15 min. Open the lid for every 5 min to avoid splitting surface due to overheat.

5. Use a brush to spread some red colouring on bun top for decoration.

純正芝麻糕
Sesame Pudding

材料

黑芝麻（炒香）120克
清水600毫升
馬蹄粉130克
清水250毫升
冰糖200克
芝麻油2茶匙

Ingredients

120g roasted black sesame
600 ml water
130g water chestnut powder
250ml water
200g rock sugar
2 tsp sesame oil

- 製作時間：**90分鐘**
- 份量：**16件**
- Production Time: **90 min**
- Yield: **16 pcs**

做法

1. 黑芝麻用篩清洗乾淨，去除沙石，隔乾水，用白鑊以中慢火兜炒至香透（約8分鐘）。
2. 芝麻加半份清水同放攪拌機內先攪拌至幼滑，加入其餘清水攪拌，用篩過濾，隔去芝麻渣。
3. 馬蹄粉與清水250毫升調成粉漿，用篩過濾。
4. 將冰糖、麻油和芝麻漿煮溶，煮時要不時攪動，離火，即時注入馬蹄粉漿內拌成稀糊狀。
5. 把芝麻漿倒入一個20厘米已掃油糕盆內，以大火蒸約60分鐘，凍透後便可切件享用。

Method

1. Use a sieve to rinse black sesame to remove dirt and drain well. Toast on a dry wok without oil by medium heat for 8 min until fragrant.
2. Put sesame and half portion of water into a blender and blend until smooth. Add in balance water then blend and sift.
3. Mix water chestnut powder and 250 ml water to batter and sift.
4. Mix and heat rock sugar, sesame oil and sesame paste until the sugar dissolved, stir occasionally. Remove from heat and stir in water chestnut batter and mix to combine.
5. Pour the mixture into on a 20 cm greased cake pan and steam over high heat for 60 min. Slice after cool and serve.

⟍ Lilian 貼心小語

市面吃到的芝麻糊很多已是採用芝麻糊粉煮成，想吃到生磨芝麻糊已不容易，何況是生磨芝麻糕？我知道有一間位於深水埗地鐵站附近的舊式士多，卻以賣傳統糕點揚名，它標榜是以生磨芝麻漿做成糕點，口感香濃軟滑，若參照我這個份量做法，既簡單，芝麻味更加香濃，只要注意清洗及炒香芝麻的方法便可。

Sesame sweet soup in market is normally made of sesame powder and it is difficult to taste freshly blended one and hence sesame cake. I know that there is a store in Shamshuipo which is famous for its conventional steaming cakes especially the sesame cake which is made from fresh blended sesame paste which brings rich and smooth flavour. Follow my easy method enhances the sesame fragrance but pay attention to the process of rinse and roasting.

奶黃馬拉糕
Custard Malay Cake

材料

麵粉 150 克，奶粉 15 克
粟粉 15 克，吉士粉 15 克
砂糖 90 克，花奶 120 克
煉奶 60 克，雞蛋 3 個
梳打粉 ¾ 茶匙
泡打粉 2 茶匙
清水 2 湯匙，牛油溶液 60 克
粟米油 30 克

奶黃餡

雞蛋 1 個，砂糖 30 克
吉士粉 20 克，麵粉 20 克
奶粉 20 克，滾水 250 毫升
煉奶（後下）80 克
熟鹹蛋黃（切碎）3 個

Ingredients

150g flour, 15g milk powder
15g corn starch, 15g custard powder
90g sugar, 120g evaporated milk
60g condensed milk, 3 eggs
¾ tsp baking soda
2 tsp baking powder
2 tbsp water, 60g melted butter
30g corn oil

Custard Filling

1 egg, 30g sugar
20g custard powder, 20g flour
20g milk powder
250 ml boiling water
80g condensed milk
3 cooked salted duck egg yolks, diced

■ 製作時間：**60 分鐘**
■ 份量：**16 件**
■ Production Time: **60 min**
■ Yield: **16 pcs**

↘ Lilian 貼心小語

據聞馥苑酒樓是最早有這款馬拉糕的酒家，大受歡迎，當然啦！我都是其中粉絲，可是這款甜點製作複雜，經我研究下，簡化過程，效果也不錯，如果要再簡單，可改為蒸小杯馬拉盞，免除了做奶黃工序，便成大受歡迎馬拉糕了。

It is said that this cake is originated from originated from Fook Yuen Seafood Restaurant and is welcome by public. I do love to taste it, however, the making method is complicated. After examining, I simplify the making steps and the result is satisfied. If you still think the process is not simple enough, you can skip the egg custard portion and turn this dessert into popular cup malay cake.

做法

1. 將麵粉、奶粉、粟粉和吉士粉同篩大碗中，加入砂糖、花奶、煉奶及雞蛋用打蛋器快速攪拌 5 分鐘至幼滑，待置 30 分鐘。

2. 梳打粉及泡打粉用清水開勻，加入粉漿內拌勻，再加入牛油熔液拌勻，分成三等份。

3. 把一份粉漿注入已墊牛油紙及掃油的長方盆內，隔水大火蒸約 10 分鐘，取出，再完成其餘兩份粉漿。

4. **奶黃餡**：把奶黃料放大碗中，用蛋拂打至幼滑，把滾水撞入蛋漿內拌勻，回小煲內煮熟，煮時要不時攪拌，離火，拌入煉奶及鹹蛋黃粒。

5. 奶黃餡趁熱塗勻在兩片馬拉糕上，疊起成三層，待涼透，便可修切成件，進食前可再翻熱享用。

Method

1. Sift flour, milk powder, corn starch and custard powder into a big bowl. Add in sugar, evaporated milk, condensed milk and egg, whisk by fast speed by whisk for 5 min until smooth. Set aside for 30 min.

2. Dissolve baking soda and baking powder in water, add in flour batter to even. Stir in melted butter to combine and divide into 3 portions.

3. Pour 1 portion of batter to greased tray which is cover by baking paper. Steam over water by high heat for 10 min. Remove and repeat the steps for balance 2 portions of batter.

4. **Egg custard :** put egg custard ingredients on a bowl and use a whisk to make all ingredients to smooth. Pour hot water into egg sauce and stir to combine. Remove back to the pot to boil. Stir occasionally. Remove and fold in condensed milk and salty diced egg yolks.

5. Spread custard filling evenly on two pieces of cakes whilst hot. Fold to 3 layers and allow to cool. Cut it up and re-heat before serve.

豆蓉千層糕
Mung Bean Paste Layered Cake

材料

去衣綠豆 80 克
滾水 200 毫升
砂糖 160 克
鹽 ¼ 茶匙
清水 400 毫升
椰漿 200 毫升
木薯粉 250 克
斑蘭葉汁 ⅓ 茶匙

Ingredients

80g peeled mung bean
200 ml boiling water
160g sugar
¼ tsp salt
400 ml water
200 ml coconut milk
250g tapioca starch
⅓ tsp pandan leave juice

■ 製作時間：**80分鐘**
■ 份量：**18件**
■ Production Time: **80 min**
■ Yield: **18 pcs**

↘ Lilian 貼心小語

泰式千層糕主要用斑蘭葉汁做成深淺兩種綠色，與印尼千層糕的做法差不多。印尼式多以三色分隔，泰式則加入豆蓉味道，兩款甜點都是用木薯粉及椰漿製做，質感非常煙韌軟滑。

Thai's layered cake mainly uses pandan leave juice to make 2 tones of green colour and the making method is similar to Indonesia's layered cake. Usually it is three- colour-layer for Indonesia one while the Thai one will add in mung bean flavour. Both are made up of tapioca flour and coconut milk with chewy and smooth texture.

做法

1. 綠豆洗淨後浸水 4 小時，隔去水，放淺窩內再注入滾水以大火蒸 20 分鐘，蓋好備用。
2. 糖、鹽加清水內煮溶，離火，加入椰漿拌勻。
3. 木薯粉放大碗中，加入暖椰漿水拌成幼滑粉漿，用篩過濾。
4. 倒出其中 270 毫升粉漿與豆蓉同放攪拌機內磨成豆蓉漿，用篩濾至幼滑，分成三份。
5. 將餘下粉漿加斑蘭葉汁調色，分成四份。
6. 預備一約 20 厘方糕盆掃油，先注入一份綠色粉漿，蓋好蒸 6 分鐘，注入一份豆蓉漿，再蒸 6 分鐘，如此類推，直至完成，最後一層是綠色粉漿，以中火再蒸 10 分鐘，取出，待凍透後方可切件。

Method

1. Rinse mung beans and soak in water for 4 hours. Drain and put on a shallow bowl and steam for 20 min. Cover and set aside.
2. Boil sugar, salt in water until dissolved. Remove and add in coconut milk and mix to even.
3. Place tapioca starch in a large bowl and add in warm coconut milk mixture and mix into batter. Sift.
4. Pour 270 ml coconut batter and mung beans into a blender and blend into bean paste. Sift until smooth. Divide into three portions.
5. Mix balance coconut batter with pandan leave juice. Divide into 4 portions.
6. Prepare a 20 cm greased cake pan. Fill in one portion of green batter in the pan and steam for 6 min. Pour one portion of bean paste on top and re-steam for 6 min. Repeat this step until pouring the final layer of green batter. Continue to steam for 10 min. Remove the pan and cool. Cut into pieces.

Cook's Tips 技術指導

蒸好的千層糕要凍透後方可切件，否則非常黐刀。建議用保鮮紙包着刀子直切，然後逐件用保鮮紙包裹好，方便食用。

Before cutting into pieces, make sure the cake has been cooled. Otherwise, it is difficult to cut. Suggest to use cling firm to cover knife and cut straightly. Then wrap each piece by cling film for convenient serving.

芒果黑白糯米飯

Black and White Glutinous Rice **with Mango**

材料

白糯米240克
黑糯米240克
斑蘭葉2片
滾水230毫升
椰漿200毫升
砂糖4湯匙
鹽¼茶匙
芒果3個
椰漿適量

Ingredients

240g white glutinous rice,
  soaked
240g black glutinous rice, soaked
2 pandan leaves
230 ml boiling water
200 ml coconut milk
4 tbsp sugar
¼ tsp salt
3 mango
Some coconut milk

■ 製作時間：**50分鐘**
■ 份量：**6人**
■ Production Time: **50 min**
■ Serving: **6**

↘ Lilian 貼心小語

有時到泰國菜館吃蠻飽，也忍不着來一個芒果糯米飯作甜品，原來椰汁糯米飯與芒果十分匹配，尤其是黑糯米，好味有益，加入斑蘭葉同蒸會散發出一點點幽香。

Sometimes, I cannot resist having mango with glutinous rice even I was full after having meal in Thai restaurant. As glutinous rice matches perfectly with mango, pandan leaves give this dessert a kind of special light fragrance and black glutinous rice is also a kind of delicious and healthy grain too.

做法

1. 黑、白糯米各自洗淨後浸水4小時，隔水，放蒸碟上。

2. 斑蘭葉剪段，放淺窩內，加白糯米及滾水30毫升同蒸35分鐘，棄去斑蘭葉。

3. 斑蘭葉段放淺窩內，黑糯米加滾水200毫升同蒸35分鐘，去掉斑蘭葉。

4. 椰漿、砂糖和鹽拌勻，分成兩份，把一份椰漿料放煲內加熱糯米飯略煮至入味，取出；把另一份椰漿加熱黑糯米飯同煮，取出，把黑、白糯米飯分別排放碟上。

5. 芒果開半起肉，伴椰汁糯米飯邊，可再淋上椰漿伴吃，增添美味。

Method

1. Rinse black and white glutinous rice and soak in water for 4 hours. Drain and steam over plate.

2. Chop up pandan leaves and place on a shallow bowl with white glutinous rice and 30 ml boiling water, and steam for 35 min. Remove leaves.

3. Place chopped pandan leaves on a shallow bowl with black glutinous rice and 200 ml boiling water, and steam for 35 min. Remove leaves.

4. Mix coconut milk, sugar and salt, divide into 2 portions. Put one portion into a pot and cooked with white glutinous rice for absorbing flavour. Dish up. Repeat up the same step to cook black glutinous rice. Then dish up on the same plate and set next to white glutinous rice.

5. Cut the mango into half, take the flesh and set next to rice. Serve with coconut milk to enhance the fragrance.

Cook's Tips 技術指導

1. 上下兩層同蒸糯米飯可省時間。

2. 蒸飯期間最好將飯翻動一次，蒸好後稍焗片刻令米飯熟透。

3. 入過冰箱的糯米飯要翻熱後才可享用。

1. It can be saved time to steam both kinds of rice at the same time by putting them into two different layers.

2. Whilst steaming, it is better to stir once and leave it covered by lid for a while after steaming. This helps rice fully cook.

3. Glutinous rice should be reheated after kept in fridge.

67

麻蓉湯丸

Sesame Rice Balls with **Sweet Ginger Soup**

材料

湯丸
糯米粉 200 克，滾水 40 毫升
清水 90 毫升，糯米粉適量

餡料
黑芝麻（炒香）60 克，糕粉 1 湯匙
糖霜 60 克，豬油 50 克

糖水料
清水 600 毫升
片糖 1 塊
薑 4 片

Ingredients

Rice balls
200g glutinous rice powder
40 ml boiling water, 90 ml water
Some glutinous rice powder

Filling
60g black sesame, roasted
1 tbsp fried glutinous rice powder
60g icing sugar, 50g lard

Sweet soup
600 ml water
1 piece of slab brown sugar
4 slice of ginger

- 製作時間：**60 分鐘**
- 份量：**8 人**
- Production Time: **60 min**
- Serving: **8**

 Lilian 貼心小語

做法

餡料：黑芝麻放機內磨成幼末，再加其餘材料拌勻，分成兩份，用膠紙包成長條，置冰箱內雪至硬身，取出，分成小粒子，再雪硬備用。

糖水：片糖加薑、清水同煮 10 分鐘至出味。

湯丸：

1. 將 ¼ 份糯米粉放大碗中，撞入滾水拌成熟粉糰，加入其餘清水及糯米粉拌均成軟滑粉糰，另加適量糯米粉搓至不沾手為止，用布蓋好備用。
2. 把粉糰搓長，分出分子，每份壓扁，包入一粒黑麻蓉，搓圓成丸子。
3. 燒大煲滾水，下丸子以中慢火煮至浮起，再滾片刻至熟透，隔起盛碗中，加入熱糖水伴食。

Method

Filling : blend black sesame to fine, add in balance ingredients and mix to even. Divide into 2 portions and wrap into roll by cling firm. Chill in the fridge until hardened. Remove and cut into small cubes. Re-chill again.

Ginger soup : Boil slab sugar, ginger and water for 10 min.

Rice balls:

1. Put ¼ glutinous rice powder in a large bow. Pour boiling water in it and stir to form dough. Add in balance water and glutinous rice powder to form soft dough. Sprinkle some glutinous rice powder and knead until clear. Use cloth to cover and set aside.
2. Roll the dough then cut into pieces and press flat. Stuff one cube of sesame paste and knead into ball.
3. Bring water to boil and put all balls in it. Boil by medium heat until floated. Keep on boiling for a while until cooked. Drain and remove on a bowl. Serve with hot ginger sugar soup.

Cook's Tips 技術指導

湯丸做多了，可存放冰格，隨時取出放滾水內煮熟便可享用。

Keep excess rice balls in freezer and re-boil before serve straightly. Then wrap each piece by cling film for convenient serving.

蓮子核桃露

Loutus Seeds and **Walnut Sweet Soup**

材料

白蓮子90克
核桃肉240克
米（浸透）60克
清水1500毫升
冰糖150克
花奶90毫升

Ingredients

90g lotus seeds
240g walnut
60g rice, soaked
1500 ml water
150g rock sugar
90 ml evaporated milk

- 製作時間：**30分鐘**
- 份量：**8人**
- Production Time: **30 min**
- Serving: **8**

做法

1. 白蓮子沖淨，先用大煲水煲15分鐘，熄火蓋好焗至水涼，取出，用牙籤穿過小孔便可把蓮芯去除，備用。
2. 核桃肉放半煲滾水內煮滾，即時撈起，吸乾水，放暖油中炸至淺金黃色，用紙吸乾油份。
3. 米浸水約4小時，隔水，與核桃肉及⅓份清水放攪拌機內磨至滑淨。
4. 餘下清水放煲內加蓮子、冰糖以慢火煲10分鐘至蓮子鬆軟，熄火，徐徐拌入核桃漿，再以中火煮核桃糊至稠身，煮時要不時攪動。
5. 最後拌入花奶，試味後便可上碗。

Method

1. Rinse lotus seeds and bring to boil for 15 min. Turn off the heat with lid covered until cool. Remove and use toothpick to push through the middle part and get rid the core. Set aside.
2. Put walnut in half pot of water and bring to boil. Remove, drain it well and deep fry in warm oil until light yellow. Absorb excess oil.
3. Soak rice in water for 4 hours. Drain and put it in the blender for blending with walnut and one third of water until smooth.
4. Bring balance water, lotus seeds, rock sugar to boil by medium heat for 10 min until lotus seeds become softened. Turn off the heat and fold in walnut paste. Boil by medium heat until thickened. Stir occasionally.
5. Add in evaporated milk. Serve.

Lilian 貼心小語

位於西環正街一間歷史悠久糖水舖（源記），它的招牌糖水如杏仁露、蓮子核桃露、桑寄生蓮子蛋茶、芝麻糊等，直到幾十年後的今天，還有不少捧場客，尤其是蓮子的鬆化可口，贏盡口碑。其實一間食肆之成功，最重要是用料上乘，再加上用心思泡製，保持水準，食客便會回味無窮。在家中煮美味糖水並不困難，只要有好材料和正確份量指引，連源記品質的蓮子核桃露也可做到呢！

There is an old sweet soup ship located in Sai Wan, famous for its almond cream, lotus seed and walnut cream, loranthaceae egg tea and sesame cream. After several decades, there are still many fans craved for its lotus seed soup. The truth is the success of one business relies on the quality of ingredients chosen and hearty effort. This will keep the heart of the customers. Actually it is not difficult to home-made delicious sweet soup, just follow instruction and choose fresh ingredients.

桑寄生蓮子蛋茶

Jishengcha with Lotus Seeds and **Egg**

材料

桑寄生200克
白蓮子120克
片糖200克
清水3.5公升
熟雞蛋8個

Ingredients

200g Jishengcha (loranthaceae)
120g lotus seeds
200g slab sugar
3.5L water
8 boiled eggs

- 製作時間：**80分鐘**
- 份量：**8人**
- Production Time: **80 min**
- Serving: **8**

Lilian 貼心小語

提到桑寄生蓮子蛋茶便會聯想起西環那間老字號甜品專家（源記），它的茶味特別香濃，清甜而不苦澀，雞蛋浸茶非常入味，蓮子鬆化而獲一致好評。最近我再次到該店品嚐，從中學習，現已掌握到烹煮方法，更知道桑寄生多益處。提議大家不妨一試，只要花十多元便可在中藥材舖買到一大盒，可分三次使用，非常經濟。

As mentioned at the previous lotus seeds and walnut cream, this tea reminds me to think of that old sweet soup shop at Sheungwan. It's tastes rich, sweet and the eggs absorb the unique flavour and is welcome by public. I have been there again to taste this tea and practice, and now I can grasp the way to make and note that this tea helps health. So why not to do it yourself?

做法

1. 白蓮子沖淨，放大煲清水內先煲15分鐘，熄火，蓋好焗15分鐘，用牙籤穿過，把蓮子芯推出，備用。
2. 桑寄生沖洗乾淨，隔水，連同3.5公升清水放煲內以慢火同煲20分鐘。
3. 雞蛋去殼，放桑寄生水內同煲40分鐘。
4. 取出雞蛋，隔去桑寄生枝；桑寄生水連同蓮子、雞蛋、片糖放回煲內再煲30分鐘，熄火，最好讓雞蛋浸入味後品嚐。

Method

1. Rinse lotus seeds and put into a pot of water for boil for 15 min. Turn off the heat and cover by lid for 15 min. Remove and use toothpick to push the core of the lotus seeds. Set aside.
2. Rinse the Jishengcha and drain well. Put it in a pot of 3.5 liters water and boil by medium heat for 20 min.
3. Remove the eggs shell and put the eggs into the Jishengcha liquid for boiling for 40 min.
4. Remove the eggs and sift the Jishengcha. Again re-boil the Jishengcha, lotus seeds and slab sugar for 30 min. Turn off the heat and let the eggs to soak further and serve.

馬來喳咋
Malay **Cha Cha**

材料

清水3.5公升
粗粒芋頭300克
三角豆80克
紅豆120克
紅腰豆120克
眉豆100克
西米60克
片糖200克
椰漿適量
花奶適量

Ingredients

3.5 liters water
300g yam, cubed
80g chick peas
120g red beans
120g kidney peas
100g black-eyed beans
60g sago
200g slab sugar
Some coconut milk
Some evaporated milk

- 製作時間：**100分鐘**
- 份量：**10人**
- Production Time: **100 min**
- Serving: **10**

⬊ Lilian 貼心小語

馬來喳咋比起紅豆沙味道及營養價值都豐富得多，濃厚的豆香味，尤其是紅腰豆及眉豆的鬆化軟腍，再加入椰漿，帶有馬來風味。我曾到油麻地"鵝記"，那是一間以喳咋盛名的糖水店，我喜歡它那粉綿綿的芋頭粒，令人回味。

The flavour and nutrient of Malay cha cha is richer than red bean soup. Intensive flavour of beans and softness of kidney peas and black-eyed beans together with the coconut milk bring exotic Malay style. You can taste this sweet soup in a local sweet soup shop located in Yaumatei. For me, I love its smooth texture of yam.

做法

1. 紅腰豆及眉豆浸水半小時。
2. 先將三角豆及紅豆放大煲水內煮滾，改用中慢火先煲30分鐘，加入芋頭粒及紅腰豆再煲30分鐘，然後加眉豆繼續煲30分鐘直至全部豆變鬆軟。
3. 西米浸水5分鐘，加入喳咋內拌勻，蓋好煮10分鐘至半透明。
4. 加入片糖煮溶，試味後便可上碗，可加入椰漿及淡奶伴食，冷熱吃均可。

Method

1. Soak kidney beans and black-eyed beans in water for half an hour.
2. Put chick peas and red beans in a pot of water and bring to boil, turn to medium heat and boil for 30 min. Add in cubed yam and kidney peas and boil for 30 min. Then add in black-eyed beans to boil for 30 min until all beans are softened.
3. Soak sage in water 5 min and put into cha cha and sitr. Cover by lid for 10 min to make semi-transparent sago.
4. Keep boiling and add in slab sugar until dissolved. Taste the sweetness. This sweet soup can be served with coconut milk and evaporated milk. Serve hot or cold.

Cook's Tips 技術指導

有人會把全部雜豆一起煲腍便算，我建議分次序下豆，做出的喳咋，豆既鬆軟又完整。

Some people will put all the beans and boil at one time. But I suggest putting them in sequences and the beans to be in complete form and softened.

蛋白杏仁茶
Egg White and **Almond Sweet Soup**

材料
大粒杏仁150克
北杏60克
清水1公升
砂糖80克
鮮奶60毫升
蛋白3個

Ingredients
150g almonds
60g bitter almonds
1 liter water
80g sugar
60 ml milk
3 egg white

■ 製作時間：**20分鐘**
■ 份量：**5人**
■ Production Time: **20 min**
■ Serving: **5**

做法

1. 杏仁、北杏沖淨後浸水1小時，隔乾，加半份清水放攪拌機內磨至幼滑，再加餘下清水再磨至細滑，倒出，用白布搾出杏仁汁，杏仁渣棄用。
2. 蛋白輕輕拌勻，備用。
3. 杏仁汁放煲內以中火煮熟，煮時要不時攪拌，加入糖及奶煮滾，熄火。
4. 徐徐拌入蛋白至稠身，上碗趁熱品嚐。

Method

1. Wash almonds and bitter almonds and soak in water for an hour. Drain well, add half portion of water and put into the blender for blending until smooth. Add in balance water and blend again until smooth. Pour into a container and use a cloth to squeeze the almond out and discard the almond ingredients.
2. Light beat the egg white. Set aside.
3. Boil the almond liquid by medium heat until cooked. Stir occasionally. Add in sugar and milk until boiled. Turn off the heat.
4. Stir in white egg until thickened. Remove and serve hot.

◥ Lilian 貼心小語

杏仁有止咳潤肺功效，杏仁茶更被視為養顏補品，要做香滑的杏仁茶，用布隔渣比用篩過濾滑溜得多，但成本較高。從前我會用粟粉水來調校稀稠，最近，我在美心酒樓品嚐過的杏仁茶很雪白又滑溜，原來他們加了蛋白，效果很好。

Almond can soothe coughing and is regarded as kind of supplement for beauty. How to make smooth almond tea? Sifting by cloth is better than sieve but the costing is higher. I have tried corn starch to make the tea thickened but the one made by Maxim Restaurant is snowy in colour and tastes smooth. The secret is they use egg white for the thickening and the effect is very good.

Cook's Tips 技術指導

自己煮蛋白杏仁茶，緊記下蛋白時不可煮成硬塊，應待杏仁茶滾起離火即拌入蛋白，蛋白不會太生也不過熟，這樣才有滑溜溜的口感。

When you make egg white and almond tea, do not make egg white lump. When the almond tea is boiled, remove and it is time to stir in egg white. This will bring silky texture.

清心丸馬蹄綠豆爽
Water Chestnuts and **Mung Beans Sweet Soup**

材料
去衣綠豆 160 克
清心丸 120 克
馬蹄肉 6 粒
冰糖 150 克
滾水 1 公升
馬蹄粉 2 湯匙
清水 100 毫升

清心丸料
生粉 80 克
滾水 150 毫升

Ingredients
160g peeled mung beans
120g qingxin balls
6 pcs of water chestnuts
150g rock sugar
1 liter boiling water
2 tbsp water chestnut powder
100 ml water

Qingxin Balls
80g corn starch
150 ml boiling water

■ 製作時間：**60分鐘**
■ 份量：**6人**
■ Production Time: **60 min**
■ Serving: **6**

Lilian 貼心小語

清心丸綠豆爽是夏日消暑最佳潮式甜品，很多人喜歡去九龍城合成甜品店品嚐，貪其夠地道。

This sweet soup is a typical Chiu Chow's summer dessert. Many people like to taste it in a sweet soup ship located in Kowloon City.

做法
1. **清心丸**：將滾水撞入生粉中，攪拌成熟粉糰，再力搓至滑淨，以木棍輾平，分切成小粒，灑上生粉撈勻，便成自製清心丸，備用。
2. 綠豆清洗乾淨，浸水 1 小時，隔乾水，放淺碟內隔水蒸 15 分鐘，備用。
3. 馬蹄肉切片；馬蹄粉用清水開成稀漿，再用粉篩濾過。
4. 冰糖放滾水內煮溶，加入綠豆、清心丸及馬蹄片滾 5 分鐘，徐徐注入馬蹄粉水不停攪拌，滾起成稀糊狀，上碗，冷熱享用均可。

Method
1. **Qingxin Ball:** pour boiling water into corn starch and stir as cooked dough. Knead until clear and roll by a rolling pin until flat. Divide it into small cubes. Sprinkle with corn starch and become qingxin balls. Set aside.
2. Rinse mung beans and soak in water for an hour. Drain well. Put on a plate and steam over water for 15 min. Set aside.
3. Cut water chestnut into pieces. Dissolve water chestnut powder into water and sift.
4. Boil rock sugar in water until dissolved and add in mung beans, qingxin balls and water chestnut for further boiling for 5 min. Fold in water chestnut liquid and stir until thickened. Remove and serve cold or hot.

Cook's Tips 技術指導

1. 清心丸可在潮州雜貨店買到。
2. 據聞清心丸用特別薯粉製成，經煮熟後晶瑩剔透，口感煙韌，但有些供應商延長保存期，會加入硼砂粉，多吃無益。自己製造，方法簡單，雖然比不上買回來的柔韌，但勝在夠新鮮健康。

1. Qingxin balls is available in Chiu Chowese grocery shops.
2. It is said that this balls is made up of tapioca flour which turns to transparent after cooked and with chewy texture. However, some suppliers will add in borax, which is harmful to health, for prolonging the expiry. If you make it yourself, the way is easy and more healthier although the texture is less chewy.

腐竹白果雞蛋糖水

Bean Curd Sheets with Gingko and Eggs **Sweet Soup**

材料
白果肉（去殼）150克
洋薏米60克
腐竹120克
滾水4公升
冰糖200克
去殼熟雞蛋6個

Ingredients
150g gingko, remove shell
60g barley
120g bean curd sheet
4 liters of boiling water
200g rock sugar
6 cooked egg, remove shell

■ 製作時間：**80分鐘**
■ 份量：**6人**
■ Production Time: **80 min**
■ Serving: **6**

↘ Lilian 貼心小語

前些日子我路經佐敦一間舊
式糖水舖，被婆婆拍白果殼
的聲音，吸引駐足坐下，享
用那碗真材實料的美味糖
水，仿如當年媽媽煮的，回
味無窮。探索追源，腐竹白
果糖水是七、八十年代家喻
戶曉的糖水，猶記當年戶外
宿營，這是熱門消夜甜品。

Not long age, I passed
though an old sweet soup
shop and there was a
granny who was removing
the shell of gingko. That
reminded me the one made
by my mom. Actually this
was a very popular dessert
after dinner in the 70-80's.

做法

1. 白果肉放少量滾水內先煮5分鐘，趁熱撕去外衣，開半去芯；洋薏米沖水備用。

2. 燒大量滾水，把白果肉及洋薏米放煲內煮滾，以中火先煮30分鐘。

3. 腐竹撕碎洗淨，加入白果水內繼續煮15分鐘，煮時要把煲蓋移開少許以防溢瀉。

4. 冰糖沖淨，與熟雞蛋加入腐竹水內再煲15分鐘至腐竹完全軟化，試味後便可享用。

Method

1. Boil gingko with some water for 5 min. Peel the skin while hot, cut half and then remove the core. Rinse the barley and set aside.

2. Boil a pot of water, put gingko and barley in it and bring to boil. Turn the heat to medium and boil for 30 min.

3. Tear bean curd sheets in shreds and rinse. Add into the gingko soup for further boiling about 15 min. Do not covered by lid completely whilst boiling and remember to leave some space to avoid overflowing.

4. Rinse rock sugar and put into the bean curd soup with eggs. Boil again for 15 min until bean curd dissolved. Test the sweetness and serve.

Cook's Tips 技術指導

1. 現代人怕剝白果殼，改用去殼白果，味道當然不夠香濃，但十分方便；也可選擇真空包裝鮮白果肉，當然也就不夠新鮮了。

2. 腐竹要揀帶有光澤及脆身的，否則會煮不溶化。

1. People do not like to get rid the shell of gingko and prefer to buy the one without shell. The flavour is certainly less rich but convenient. Vacumn packed fresh ginko is more convenient, but the quality is not ensured.

2. Choose those bean curd sheet with shine and crispy, otherwise, it can not be melted while boiling.

參鬚元肉粟米水
Ginseng Root with Longan and **Corn Soup**

- 製作時間：**30分鐘**
- 份量：**6人**
- Production Time: **30 min**
- Serving: **6**

材料
參鬚40克，粟米2條
元肉60克，清水3公升
冰糖100-150克

Ingredients
40g ginseng root, 2 corns
60g longan, 3 liters of water
100-150g rock sugar

做法
1. 所有材料沖洗乾淨同放煲。
2. 注入清水以大火煲滾，轉中火煲約30分鐘。
3. 下冰糖煲溶，試味後便可關火。
4. 隔起參鬚、粟米及元肉，可取吃或棄去。參鬚水冷熱皆可，不過雪凍會更加可口。

Method
1. Rinse all ingredients and place in a pot.
2. Pour water in the pot and bring to boil by high heat. Turn to medium heat and boil for 30 min.
3. Dissolve rock sugar into the soup. Test the sweetness and switch off the heat.
4. Discard the ingredients or serve with the soup. For the ginseng soup, serve whilst hot or cold. It will be more delicious after chilled.

↘ Mindy 貼心小語

一個看似不大吸引，起初對它的味道有點抗拒的參鬚水，冠以一個很有意思的名字"分甘同味"，它源自前港督御廚，有幸品嚐者都會對那種回甘在口的健康飲品，印象難忘。

This soup is originated from the cook of former governor with a meaningful name – "Sharing the Bitterness. Although this soup seemed unattractive initially, I then loved the after taste.

Cook's Tips 技術指導

參鬚可增強免疫力和不易上火，加入粟米及元肉後，更增添甜味。

Ginseng can enhance the immune system and corn and longan add the sweetness.

羅漢果茶
Siraitia Tea

製作時間：**20分鐘**
份量：**4人**
Production Time: **20 min**
Serving: **4**

材料
羅漢果1個，甘草5片
南北杏40克，清水2公升

Ingredients
1 siraitia, 5 pieces liquorice
40g sweet and bitter almonds,
2 liters of water

做法
1. 把羅漢果、甘草和南北杏洗淨。
2. 壓碎羅漢果連外殼及果肉，注入清水一同煲滾。
3. 轉中火煲20鐘，熄火。
4. 原煲燜焗40分鐘至羅漢果出味，這便是一道甘甜健康的羅漢果茶了。

Method
1. Rinse siraitia, liquorice and almonds.
2. Crash the whole siraitia and put into a pot and boil with water.
3. Turn to medium heat and boil for 20 min. Turn off the heat.
4. Leave the pot covered by lid for 40 min so that it brings full flavour of siraitia.

Cook's Tips 技術指導

1. 羅漢果茶無需添加糖份，充滿着甘甜的芳香，冷熱均宜。
2. 把羅漢果壓碎，容易在烹煮時滲出味道。

1. No need to add sugar as siraitia is rich with it's unique bitter sweetness. Serve cold or hot.
2. It is easier to let the flavour of siraitia come out after crushed.

Mindy 貼心小語

羅漢果的甜度是砂糖的300倍，但熱量近乎零，其清熱潤肺和防止色斑、皺紋及皮膚老化，美容妙品，因而有中華神果的美號。在桂林，它是招呼貴賓的上品。在香港，高級食府"利苑"也用此道飲品招呼來賓。

Siraitia has its unique bitter sweetness which is 300 times sweeter than normal sugar but energy is almost zero. Besides, it help improve health condition and with rejuvenating effect. In Guilin, a place in mainland, this drink is for serving special guests. In Hong Kong, a high-end chinese restaurant uses this drink to serve special guests too.

清補涼冰
Icing Ching Po Leung **Soup**

材料

淮山 30克
茨實 30克
沙參 30克
玉竹 30克
蓮子 30克
百合 30克
薏米 30克
元肉 30克
紅棗（去核）6粒
清水 4公升
冰糖 150-180克

Ingredients

30g chinese yam
30g euryale seeds
30g glehnia root
30g solomonseal
30g lotus seeds
30g lily bulb
30g barley
30g longan
6 red dates
4 liters of water
150-180g rock sugar

做法

1. 將淮山、茨實、沙參和玉竹，先用熱水浸半小時，瀝乾備用。
2. 連同其餘材料沖洗乾淨，放凍水煲滾。
3. 轉用慢火煲2小時，放冰糖煲溶後試味。
4. 待涼放冰箱，雪凍後便成美味有益的清補涼冰了。

Method

1. Rinse Chinese yam, euryale seeds, glehnia root, solomonseal and soak in hot water for half an hour. Drain and set aside.
2. Rinse balance ingredients and place all ingredients in a pot of water and boil.
3. Turn to low heat and boil for 2 hours. Add in rock sugar and test the sweetness.
4. Leave it cool and chill in the fridge.

⊠ Mindy 貼心小語

到人氣食店"火車頭越南菜館"食飯，在餐牌上看到清補涼冰，好奇之下點了這個飲品。清補涼是香港家庭常見的湯水，這道湯水有潤肺健胃的功效，一年四季都很適合飲用。這個清補涼冰也有同樣功效吧！清甜中帶點蓮子、百合的香味，夏天飲用真是冰涼又滋補，為了自己及家人，做一個既好味又有益的飲品。

This soup is a typical home-made soup in Hong Kong as it has soothing effect to stomach and lung and is a all-year-round drink. I have been to a hit Vietnam restaurant and found this drink was available but serve cold. I believe this drink will have same effects too! Actually this healthy drink is good for you and family and the flavour is full of fragrance of lotus seeds and lily bulb.

流心芒果蛋糕
Mango **Cake**

■ 製作時間：**90分鐘**
■ 模具：**7吋直徑半球形模具**
■ Production Time: **90 min**
■ Yield: **7" diameter, semi circle mould**

材料

芒果片（裝飾）適量
甜忌廉150克（打起，貼芒果片
　用）適量

清蛋糕：（林明頓的清蛋糕⅓份量）
清蛋糕（直徑7吋）1片
清蛋糕（直徑4吋）1片

流心芒果餡
芒果茸200克，芒果粒150克

芝士餡
芒果茸160克
忌廉芝士（室溫）160克
淡忌廉（打起）130克
魚膠片（冰水浸軟）8克
蛋黃（大）1個，砂糖40克
檸檬汁2茶匙

Ingredients

Some mango slices (for decoration)
150g sweet cream (whisked, for
　sticking with mango slices)

**Sponge Cake : (⅓ of Lamington
cake)**
1 slice of cake (7" diameter)
1 slice of cake (4" diameter)

Liquefied mango Filling
200g mango purée
150g mango cubes

Cheese Filling
160g mango purée
160g cream cheese (room temperature)
130g whipped cream (whisked)
8g gelatine leaf (soaked in ice water)
1 egg yolk, 40g sugar
2 tsp lemon juice

做法

流心芒果餡： 混合所有材料，放冰格雪成圓碟狀冰塊，備用。

芝士餡：

1. 已浸軟的魚膠片加入檸檬汁坐熱水至溶化，備用。
2. 蛋黃加砂糖坐滾水上，打至呈淺黃色及稠身，加入忌廉芝士內打勻，順序加入魚膠液，芒果茸及淡忌廉，全部混合。

組合：

1. 濕透模具內部，倒入⅓芝士餡，拌勻在整個球型模具中。
2. 放入直徑4吋蛋糕片，再倒入⅓芝士餡，然後放入芒果茸冰塊，再倒入餘下的芝士餡，蓋上直徑7吋清蛋糕片，用手輕輕按貼蛋糕，放冰箱3小時至凝固。
3. 取出脫模，抹上薄薄一層已打企身的甜忌廉，最後將芒果片順方向放上球體上，隨意裝飾便可。

Method

Liquefied mango filling : mix up all ingredients and chill it in the freeze to form small round ice cubes. Set aside.

Cheese Filling:

1. Put soaked gelatine leaf into lemon juice and melt over hot water. Set aside.
2. Add egg yolk into sugar and set over boiling water. Whisk until pale yellow and thickened. Add in cream cheese and mix to combine. Put gelatine liquid, mango purée and whipped cream, one by one, and stir together.

Assembly:

1. Wet the cake mould and pour ⅓ cheese filling and spread evenly on the cake mould.
2. Place the 4" cake slice and pour ⅓ cheese filling then liquefied mango filling. Add in balance cheese filling and cover by 7" cake slice. Lightly press the cake and chill it in the fridge for 3 hours until hardened.
3. Remove the mould and spread a thin layer of whisked, sweet cream on top. Stick mango slices on the surface orderly. Have some decoration finally.

Cook's Tips 技術指導

1. 流心芒果的秘訣：將芒果茸及果粒雪至結冰，放入芝士餡內，待芝士餡凝固，而芒果茸退冰，切開蛋糕便會流出。
2. 如果家裏沒有球形模，可用一個大湯碗替代。或用蛋糕模做成切餅的形狀，組合時先放蛋糕片，再倒入芝士餡，如此類推，次序與球形模相反便是了。

1. The secret of making liquefied mango is to freeze the mango purée and mango cubes and then put into the cheese filling. When cheese filling is hardened, ice of the mango purée is melted. Once the cake is cut, there comes liquefied effect.
2. You can replace the semi-spherical mould by a large bowl. Or you can use cake mould to form shape of cut cake. While assembling, just reverse the assembly sequence.

榴槤椰香慕斯
Durian and Coconut Mousse

材料

金枕頭榴槤肉200克
椰汁60克
蛋黃2個
砂糖30克
淡忌廉（打起）150克
魚膠片（冰水浸軟）4克

Ingredients

200g durian
60g coconut juice
2 egg yolk
30g sugar
150g whisked cream
4g gelatine leave (soaked in ice water)

- 製作時間：**30分鐘**
- 份量：**4杯**
- Production Time: **30 min**
- Yield: **4 cups**

Mindy 貼心小語

金枕頭又叫馬驪頭，它的外殼釘紋沒紋理，尾部尖尖，肉多核細，味道特別清香。這個榴槤椰香慕斯，做法簡單，加入椰汁能提昇它的香甜味道，沒騙你，不愛榴槤的人，也會愛上它。

Monthong, this kind of durian has distinctive unique aroma, small kernel, rich flesh and there is no special pattern for the thorn. This durian and coconut mousse is easy-to-make and delicious; and coconut milk can enhance durian's flavour. You will love this dessert after tasting.

做法

1. 椰汁加入榴槤肉用打蛋器打成糊狀，備用。
2. 蛋黃加砂糖坐滾水打至淺黃稠身，加入已浸軟的魚膠片拌溶，再加入榴槤糊混合。
3. 淡忌廉打至企身，加入榴槤糊內混合拌勻。
4. 置冰箱雪至凝固，便成又香又甜的榴槤椰香慕斯了。

Method

1. Put coconut milk into durian flesh and whisk to batter. Set aside.
2. Whisk egg and sugar over boiling water until thickened. Add in gelatine leave and mix, and fold in durian batter to combine.
3. Whisk the cream to stiff peak. Put cream into durian paste and mix together.
4. Chill the mixture in the fridge until set.

Cook's Tips 技術指導

如果喜愛做切餅，只要預備2片清蛋糕和模具，倒入慕斯，雪至凝固便可隨意裝飾。

If you like to make cut cakes, just prepare two slices of cake and mould. Then pour mousse into the mould and chill until set. Decorate and serve.

自家製熱情果 / 藍莓雪糕
Home-made Passion Fruit / **Blueberries Ice Cream**

■ 製作時間：**90分鐘**
■ 份量：**1公升**
■ Production Time: **90 min**
■ Yield: **1 Liter**

 Mindy 貼心小語

自家製雪糕甜度可調校，用新鮮水果，真忌廉，避免人造色素及香料，吃起來健康得多了，比起街上買的雪糕，自家製的成本一點也不平宜，但勝在真材實料。

You can adjust the sweetness of home-made ice cream and it is healthier to make it by using fresh fruits, cream and avoid artificial colouring and essence. However, the cost is higher but you can enjoy fresh ingredients.

材料

熱情果雪糕

熱情果汁250克，蛋黃3個，砂糖120克
全脂鮮奶300克，淡忌廉300克

藍莓雪糕

藍莓400克，砂糖45克，檸檬汁50克
清水50克，蛋黃3個，砂糖45克
全脂鮮奶300克，淡忌廉300克

Ingredients

Passionate Fruit Ice Cream

250g passionate fruit juice
3 egg yolk, 120g sugar
300g whole fat milk, 300g whipped cream

Blueberries Ice Cream

400g blueberries, 45g sugar, 50g lemon juice
50g water, 3 egg yolk, 45g sugar
300g whole fat milk, 300g whipped cream

做法

熱情果雪糕：

1. 熱情果肉過篩隔去果籽，果汁重約250克，放冰箱結冰。
2. 蛋黃、砂糖加鮮奶30克攪均。
3. 鮮奶270克煲滾離火，將一半鮮奶慢慢倒入蛋黃奶內，邊倒邊攪拌，以免蛋漿被灼熟。
4. 將蛋漿混合物倒回煲內，開中火邊煮邊攪拌，煮至濃稠，離火，過篩，放入熱情果冰，令蛋漿冷卻。
5. 淡忌廉打起加入熱情果汁蛋漿混合，倒進雪糕機內雪約30-45分鐘便完成了，如硬度不夠，放冰箱雪硬便可享用了。

藍莓雪糕：

1. 藍莓、砂糖45克、檸檬汁及清水用中火煮約10分鐘，期間不時攪拌，以免煮焦，待煮至濃調，藍莓醬汁約有250克放冰箱結冰備用。
2. 依照熱情果雪糕程序處理便可。

Method

Passion Fruit Ice Cream:

1. Sift passion fruit flesh to get rid the seeds and juice weighs around 250g. chill it in the fridge.
2. Add egg yolk, sugar to 30g milk and stir to even.
3. Bring 270g milk to boil. Remove from heat and pour half the milk into egg sauce. Stir whilst mixing so as to avoid the egg sauce being cooked.
4. Put egg sauce back to a pot and boil by medium heat. Stir occasionally until thickened. Remove from heat and sift. Put on the icy passion fruit to cool.
5. Whisk whipped cream and fold in the passion fruit mixture. Mix to combine and pour into a ice-cream machine to chill for 30-45 min. If it is not hardened enough, chill it in the fridge. Serve.

Blueberries Ice Cream:

1. Bring blueberries, sugar 45g, lemon juice and water to boil on medium heat for 10 min. Stir occasionally to avoid overcooked. The cooked thickened sauce weighs around 250g and chill in the fridge. Set aside.
2. Follow the steps of making passion fruit ice cream.

Cook's Tips 技術指導

如果沒有雪糕機，把做好的材料，放入冰箱約40-50分鐘，然後用打蛋器打鬆散呈半凝固的雪糕，再放回冰箱，重複以上步驟約4-5次見雪糕已成沙冰狀，再雪至硬身，便可享用了。

If you do not have ice cream machine, you can chill all the ready ingredients in the fridge for 40-50 min. Then use a whisk to whisk the mixture until soft and put back to the fridge. Repeat the above steps for 4-5 times, the mixture will turn to icy form. Serve after the mixture is hardened.

藍莓配忌廉芝士
Blueberries and **Cream Cheese**

■ 製作時間：**45分鐘**
■ 份量：**6杯**
■ Production Time: **45 min**
■ Yield: **6**

↘ Mindy 貼心小語

藍莓芝士蛋糕吃多了，今次我把它混入酸忌廉和檸檬汁，突顯果香。幼滑慕斯沾滿沒添加劑的自製藍莓醬，吃得健康與否，並不重要，反而是美味掛帥。

A bit bored about blueberries cheese cake? I try to add in sour cream and lemon juice to enhance its distinctive fruit aroma. How can you resist a mouthful of fine, yummy mousse with self-made fresh blueberry jam? Forget those healthy motto at the moment.

材料
酸忌廉芝士餡
忌廉芝士（室溫）200克
酸忌廉（室溫）100克
砂糖100克，清水30克，蛋黃2個
魚膠片9克，檸檬汁20克
淡忌廉（打起）200克

藍莓醬
藍莓400克，砂糖45克
檸檬汁50克，清水50克

Ingredients
Sour Cream Cheese Filling
200g cream cheese (room temperature)
100g sour cream (room temperature)
100g sugar, 30g water, 2 egg yolk
9g gelatine leave, 20g lemon juice
200g whipped cream, whisked

Blueberries Jam
400g blueberries, 45g sugar
50g lemon juice, 50g water

做法
酸忌廉芝士：
1. 魚膠片用冰水浸軟，加入檸檬汁坐熱水溶化，待用。
2. 忌廉芝士及酸忌廉混合均勻。
3. 砂糖加清水煲至大滾。一邊攪拌蛋黃，慢慢倒入滾沸糖水，打起至濃稠。
4. 把½份蛋黃漿加入魚膠粉溶液內拌勻，再倒回餘下的蛋漿內混合，分數次拌入芝士酸忌廉，最後拌入已打起的淡忌廉。

藍莓醬：藍莓、砂糖、檸檬汁及清水以中火煮約10分鐘，煮時略按壓出藍莓汁，顏色更漂亮，放涼待用。

組合：將保鮮紙墊放在小杯中，倒入半杯慕斯，放入1湯匙藍莓醬，再倒滿慕斯，扭緊如網球大小，放冰箱至凝固，吃時淋上藍莓醬。

Method
Cream Cheese:
1. Soak gelatine leaf in ice water, add in lemon juice and melt over hot water. Set aside.
2. Mix cream cheese and sour cream.
3. Bring sugar and water to boil. Stir egg yolk and lightly add in boiling water until thickened.
4. Put ½ egg yolk sauce into gelatine sauce and stir to combine. Pour into the balance egg yolk sauce and mix together. Fold in cheese mixture by several additions. Finally fold in whipped cream.

Blueberries Jam : Bring blueberries, sugar, lemon juice and water to boil by medium heat for 10 min. Lightly squeeze blueberry juice while cooking and this makes nice colour. Cool and set aside.

Assembly : Line cling firm on a small cup and pour half cup of mousse. Place 1 tbsp of blueberries jam and fill up the cup by mousse. Twist the cling firm forming tennis ball size. Chill in the fridge until hardened. Serve with blueberries jam.

Cook's Tips 技術指導

藍莓醬是這慕斯的靈魂，不妨煮多一點，留作伴雲喱拿雪糕或芝士蛋糕都可以。

Blueberries jam is the soul of this dessert. Try to make more for further serving with vanilla ice cream or cheese cake.

提拉米蘇（意大利咖啡芝士餅）
Tiramisu

■ 製作時間：**45分鐘**
■ 模具：**8吋圓形糕盆**
■ Production Time: **45 min**
■ Utensil: **8" cake pan**

↘ Mindy 貼心小語

歷久不衰的Tiramisu在香港比任何一款蛋糕更受歡迎，它可以説是把芝士蛋糕帶進我們的生活中。這道甜品傳統會用大盆子盛載，不過改做為切餅更加吸引，因為吃不完，可原件保存在冰箱，不影響到造型和食味。

Tiramisu is an everlasting cake which is more popular than any other cake. It seems a prologue for us to encounter cheese cakes. The conventional tiramisu is contained in a large bowl but it is more attractive to be transformed in cut cakes. As we can chill the left over in fridge with keeping the flavour and shape.

材料
咖啡蛋糕（直徑8吋）3片（參閱林明頓）
咖啡粉2湯匙

咖啡溶液
咖啡粉2湯匙，咖啡糖1湯匙，滾水3湯匙
咖啡酒（Kahlua）3湯匙
杏仁酒（Disaronno Amaretto）1湯匙

芝士餡
蛋白2個，砂糖50克，淡忌廉（打起）200克
蛋黃2個，Mascarpone 馬士卡峰芝士250克
魚膠粉2茶匙，凍水3湯匙，鮮奶3湯匙

裝飾
可可粉（灑面用）適量

Ingredients
3 slices 8" coffee cake (ref to Lamington)
2 tbsp instant coffee

Coffee Solution
2 tbsp instant coffee, 1 tbsp coffee sugar
3 tbsp boiling water
3 tbsp Kahlua (coffee wine)
1 tbsp Disaronno Amaretto (almond wine)

Cheese Filling
2 egg white, 50g sugar
200g whipping cream, whisked
2 egg yolk, 250g Mascarpone cheese
2 tsp gelatine powder
3 tbsp cold water, 3 tbsp milk

Granishing
Some cocoa powder, for dusting

做法
咖啡蛋糕：請參照林明頓蛋糕（第16頁）做法，只是把鮮奶加進即溶咖啡便可。
咖啡溶液：咖啡粉及咖啡糖用滾水拌溶，加酒拌勻，備用。

芝士餡：
1. 魚膠粉先用凍水拌勻，隔水坐熱至完全溶化，待用。
2. 蛋白加砂糖打起。蛋黃坐熱水打至淡黃色，加入芝士內拌至軟滑。
3. 鮮奶與魚膠粉溶液拌勻，再與¼份咖啡溶液拌入芝士內，輕輕拌入淡忌廉及已打起的蛋白。

組合：
1. 把一片蛋糕片放入模具內，掃上⅓份咖啡溶液，倒入⅓份芝士餡，重複以上步驟，直至芝士鋪滿至模邊，置冰箱待凝固最少6小時。
2. 蛋糕脫模，灑上無糖可可粉及可隨意裝飾，切件享用。

Method
Coffee Cake : refer to the method of making Lamington (p.16). Simply put milk into instant coffee.
Coffee Solution : dissolve coffee powder and coffee sugar in boiling water. Add in wine and stir to even. Set aside.

Cheese Filling:
1. Stir gelatine powder with cold water and mix together. Melt over hot water until dissolved. Set aside.
2. Whisk egg white and sugar until stiff. Whisk egg yolk over hot water until light yellow. Add in cheese and stir until smooth.
3. Mix milk and gelatine liquid and add in ¼ coffee solution together with cheese. Gently fold in whipped cream and whisked egg white.

Assembly:
1. Line one slice of cake in the cake mould. Brush ⅓ coffee solution on it. Pour in ⅓ cheese filling. Repeat the procedure until fill up the mould. Chill in fridge for 6 hrs to set
2. Remove from the mould, dust with cocoa powder. You can granish the cake as you like. Cut into pieces and serve.

Cook's Tips 技術指導

這食譜容易掌握，若想省點功夫，可買現成咖啡蛋糕；或跟傳統方法，把咖啡蛋糕改用手指餅吧！
It is easy to grasp the step of this receipe. If you want to simplify the process, just purchase a coffee spange cake. Or you can follow traditional method, replace coffee cake by finger biscuits.

荔枝奶凍
Lychee **Panna Cotta**

材料

荔枝汁200克
淡忌廉300克
鮮奶300克
砂糖70克
魚膠片（冰水浸軟）12克
荔枝肉（份量以外）隨意

Ingredients

200g lychee juice
300g whipped cream
300g milk
70g sugar
12g gelatine leaf (soaked in ice water)
some lychee flesh (optional)

- 製作時間：**20分鐘**
- 份量：**6杯**
- Production Time: **20 min**
- Yield: **6 cups**

做法

1. 荔枝肉放攪拌機打出汁液過篩，留汁約200克備用。
2. 淡忌廉、鮮奶及砂糖煲滾離火。
3. 放入已浸軟魚膠片和荔枝汁混合拌溶。坐熱水拌溶。
4. 分別倒入小杯，加入少許荔枝肉，放冰箱雪至凝固（約4小時），便可享用。

Method

1. Put lychee flesh in the blender and blend. Sift out the juice and leave 200g.
2. Bring whipped cream, milk and sugar to boil. Remove from heat.
3. Add milk mixture in gelatine leaf with lychee juice. Stir well and melt over hot water.
4. Pour the mixture into small cups and top with some lychee flesh. Chill in the fridge for 4 hours. Serve after set.

Mindy 貼心小語

近年熱賣的荔枝奶凍，聽說用上北海道3.6牛乳，更添香滑，加上名人效應，令人人趨之若鶩。自己動手製造，可以說是零難度，只要有荔枝汁便可，無論是新鮮荔枝汁或罐裝貨，效果都很好。

It is said the this recent hit lychee panna cotta is made up of 3.6 Hokkaido whole milk which brings rich and smooth effect. In addition to the promotion of celebrities, people are crazy about it. Actually, this is definitely a eashy-to-make dessert. The effect is pretty good no matter you use fresh lychee juice or canned lychee juice.

Cook's Tips 技術指導

在荔枝季節裏，用新鮮荔枝打汁最好吃，試過後保證人人歡喜。不是時令時，可用罐頭荔枝代替，隔去糖水和減少糖份，一年四季也可以大快朵頤啊！

During the period of lychee season, lychee juice is very tasty and every one loves it. If no lychee is available in market, using canned one but drains out the syrup and reduces the sugar quantity. Then, you can make this dessert all year round.

香濃芒果布甸
Mango **Pudding**

材料
芒果味啫喱粉1盒
魚膠粉10克
砂糖30克
滾水250毫升
冰水250毫升
凍淡奶250毫升
芒果味雪糕100克
大熟芒果2個

Ingredients
1 box of mango flavour jelly
  powder
10g gelatine powder
30g sugar
250 ml boiling water
250 ml ice water
250 ml cold evaporated milk
100g mango ice cream
2 ripe mangoes

■ 製作時間：**30分鐘**
■ 份量：**8杯**
■ Production Time: **30 min**
■ Yield: **8 cups**

Lilian 貼心小語

芒果布甸一直大受歡迎，不論餅店或酒樓都有出品，但味道及顏色有點過於人造化。我說甜品屋如（許留山）或自家製的布甸才可滿足愛芒果人士，我這布甸還加入了雪糕，這令布甸更加香滑及色澤自然，更吸引。

This mango pudding has been very popular for years and you can find it in cake shops or even in the restaurants. However, the flavour and the colour of some of the products seem a bit artificial. The pudding quality made by a local dessert soup, named Hui Lau Shan, or DIY types can only satifsy mango lovers. Mine add in ice cream which enhance the smoothness and colour.

做法
1. 將啫喱粉、魚膠粉及糖放入袋中搖勻，然後放入煲中，倒入滾水中拌至完全溶解，離火，待涼透。
2. 芒果去皮，把半份果肉切粒，其餘磨成果蓉，備用。
3. 凍啫喱水加冰水拌勻，再拌入凍淡奶、雪糕、果蓉及芒果粒至半凝固狀。
4. 立即倒入已灑過水之啫喱模型中，入冰箱雪至凝固，便可享用。

Method
1. Put jelly powder, gelatine powder and sugar in a bag and shake to even. Place it in a pot of boiling water and mix until dissolved. Remove from heat and cool.
2. Skin the mangoes and cut half of the flesh into dices. Make purée for the balance. Set aside.
3. Add jelly liquid with ice water and mix to combine. Fold in cold evaporated milk, ice cream, mango purée and mango dices until semi-hardened.
4. Pour the mixture into jelly mould which is pre-coated with water. Chill it in the fridge and serve.

木糠布甸
Serradura

製作時間：**45分鐘**
份量：**6杯**
Production Time: **45 min**
Yield: **6 cups**

◥ Mindy 貼心小語

第一次吃木糠布甸在澳門，初嚐感覺，驚為天人，但自從香港開了分店售賣各款味道的木糠布甸後，不知是否改了食材，味道總是不同了！如果你也愛木糠布甸，便試試這個食譜吧，味道蠻不錯！這次我把布甸裝飾像小盆栽，看看誰有膽量連泥土也一起吃掉！

My first experience to taste Serradur was in Macau. It was so amazing about the taste of it. But the taste seems different when I tried it again in local branches. If you do like this dessert, follow this receipe and make one for yourself and the taste is quite good. See, I decorate my Serradura as a small plant and see if you are brave enough to eat those soil !

材料

木糠料

Oreo餅乾70克，可可粉35克

布甸料

蛋黃4個，砂糖40克，鮮奶125克，魚膠粉5克
清水3湯匙，淡忌廉500克，煉奶100克
雲呢嗱油½茶匙

Ingredients
Serradura
70g Oreo biscuits, 35g cocoa powder

Pudding
4 egg yolks, 40g sugar, 125g milk
5g gelatine powder, 3 tbsp water
500g whipped cream, 100g condensed milk
½ tsp vanilla essence

做法

木糠： Oreo餅用木棍壓碎或用攪拌器打碎，拌入可可粉備用。

布甸：

1. 蛋黃加砂糖拌勻，將鮮奶煲至微滾，取半份倒入蛋黃中，一邊倒，一邊用打蛋器把蛋黃攪拌，以免熱奶將蛋黃燙熟，倒回煲內，以中慢火加熱及攪拌，直至濃稠為止，加入雲呢嗱油拌勻，待涼。
2. 魚膠粉加水拌勻，用熱水坐溶。
3. 淡忌廉加煉奶用攪拌器打至八、九成企身。
4. 將魚膠粉水加入蛋黃漿中完全拌勻，將⅓淡忌廉加入拌勻，然後將餘下的淡忌廉加入攪拌至完全混合為止。

組合： 先將木糠料放入蛋糕模內，然後將布甸放入唧袋中打圈唧一層，再加上木糠料，如此一層一層鋪上，放入冰箱的冷藏室，或放冰箱雪成布甸狀。兩種吃法截然不同。

Method

Serradura : use a rolling pin crush the Oreo or blend by a food blender. Fold in cocoa powder and set aside.

Pudding

1. Mix egg yolk and sugar. Bring milk to a boil and pour half portion into egg yolk. Stir with a whisk while pouring so as to avoid the egg yolk is cooked by hot milk. Pour the mixture into the pot and boil by medium heat. Stir occasionally until thickened. Add in vanilla essence. Cool.
2. Mix gelatine powder and water. Melt over hot water.
3. Whisk whipped cream and condensed milk by a mixing bowl until almost stiff peak.
4. Stir gelatine liquid with egg yolk mixture until even, then fold in ⅓ whipped cream and combine. Fold in balance whipped cream until all mix together.

Assembly : put the Serradura into the cake mould. Put the pudding ingredients in a piping bag and squeeze a round layer. Top with Serradura again and a layer of pudding and so on. Chill in the freezer until hardened as ice cream or set in fridge as the hardness of pudding.

Cook's Tips 技術指導

1. 原味的木糠布甸用瑪莉餅。
2. 若加入榛子粉更香口，但個人最愛Oreo餅加可可粉。
1. Conventinal Serradura is made of Mary biscuits.
2. If add in hazelnut powder, it can enhance the fragrance. But I personally love to use Oreo biscuits and cocoa powder.

鴛鴦咖啡糕
YuanYang **Coffee Pudding**

■ 製作時間：**100分鐘**
■ 份量：**10人**
■ Production Time: **100 min**
■ Serving: **10**

材料

忌廉咖啡凍
特濃咖啡1湯匙，滾水50毫升
魚膠粉18克，冷開水60毫升
蛋黃3個，砂糖75克
鮮奶270毫升，淡忌廉150毫升

咖啡凍
魚膠粉25克，冷開水70毫升
滾水450毫升，砂糖150克
特濃咖啡3湯匙

Ingredients
Cream Coffee pudding
1 tbsp espresso coffee
50 ml boiling water
18g gelatine powder
60 ml cold boiled water
3 egg yolk, 75g sugar, 270 ml milk
150 ml whipped cream

Coffee pudding
25g gelatine powder
70 ml cold boiled water
450 ml boiling water, 150g sugar
3 tbsp espresso coffee

◢ Lilian 貼心小語

我在酒樓吃過的鴛鴦咖啡糕也不及自己做的香濃，因自己可選用特濃咖啡，愛咖啡人士會更加滿足。

The taste of the Yuan Yang coffee pudding I made is richer than the one I tried in a famous Restaurant. The tip is the adding of espresso. Coffee lovers surely will be satisfied by this flavour.

做法

1. **咖啡凍**：魚膠粉放碗內加水開勻；燒滾水，加糖煮溶，再加咖啡及魚膠粉水拌至完全溶解，待涼。

2. **忌廉咖啡凍**：咖啡、魚膠粉分別用水開溶；蛋黃加糖攪拌至淡黃色，鮮奶放小煲內煮熱，撞入蛋黃中，攪勻，再回煲內煮滾，煮時要用蛋拂不停攪拌，加魚膠粉水及咖啡溶液拌勻，待涼。

3. 忌廉打起至稠身，拌入以上凍蛋漿內，備用。

4. 預備已灑水的方形模，先注入⅕忌廉咖啡凍，置冰箱內待凝固，取出，再注入⅕份咖啡凍，待凝固，如此類推，直至完成最後一層咖啡凍為止，入冰箱中冷凍至完全凝固，便可切件享用。

Method

1. **Coffee pudding :** dissolve gelatine powder in water. Bring water to boil and add in sugar until dissolved. Pour coffee powder and gelatine liquid in sugar liquid and stir until dissolved. Cool.

2. **Cream Coffee pudding :** dissolve coffee, gelatine powder with water separately. Whisk egg yolk and sugar until light yellow. Pour milk in a pot and boil, remove and add in egg yolk. Stir to combine. Put back in the pot and boil. Stir by a whisk occasionally. Add in gelatine liquid and coffee solution and mix to even. Cool.

3. Whisk whipped cream until thickened. Fold in egg batter. Set aside.

4. Sprinkle some water on a square cake mould and pour ⅕ of the cream coffee batter. Chill in the fridge until solidified. Remove and pour ⅕ coffee batter, chill it in the fridge until hardened. Repeat the above steps until accomplishing the final layer of coffee batter. Chill it until hardened. cut into pie before serve.

Cook's Tips 技術指導

1. 注意時間掌握，若咖啡凍未曾凝固便注入第二層液體，會令鴛鴦顏色混合，層次不夠分明。

2. 容器的容量要合適，太大的容器要改做八層，否則每層厚度不平均。

1. Pay attention to the time control. If the coffee jelly has not been hardened completely, down pouring of second layer will stain the layer beneath and affect the colour finishing.

2. Choose appropriate mould, change to do eight layers if the mould is too large. Otherwise, the thickness of each layer will be uneven.

咖啡奶凍
Coffee **Panna Cotta**

- 製作時間：**30分鐘**
- 份量：**6-8杯**
- Production Time: **30 min**
- Yield: **6-8 cups**

↘ Mindy 貼心小語

曾幾何時，放學後不是立刻回家，而是到超級市場購買一杯一杯小小的咖啡布甸，那杯小布甸滑不溜口，表面沾滿焦糖漿，但不知甚麼原因，這款布甸現在已經銷聲匿跡，要回味便要自己動手了。現在把咖啡配上雲喱拿，伴着焦糖脆脆一層一層吃下去，又另一番風味，想起也流口水啊！

Somewhere in time, I dashed forward to buy a small cup of coffee pudding after school but not going back to home. I really miss those fine small puddings coated with caramel. However, this kind of dessert is no longer exist in the market. I have to make it myself when I want to taste it. Image layers of coffee pudding and vanilla pudding with crispy caramel, I really want have a bite now !

材料

焦糖脆脆（放面）伴吃

咖啡奶凍
淡忌廉300克，鮮奶100克，咖啡1 ½ 湯匙
咖啡糖45克，咖啡酒10克，魚膠片（冰水浸軟）4克

雲喱拿奶凍
淡忌廉300克，鮮奶100克，砂糖45克
雲喱拿豆½ 條，魚膠片（冰水浸軟）4克
杏仁酒10克

Ingredients

Crispy caramel (for topping)

Coffee Panna Cotta
300g whipped cream, 100g milk
1 ½ tbsp coffee, 45g coffee sugar
10g Kahlua (coffee wine)
4g gelatine leaf (soaked in ice water)

Vanilla Panna Cotta
300g whipped cream, 100g milk, 45g sugar
½ vanilla pot, 4g gelatine leaf (soaked in ice water)
10g Disaronno Amaretto (almond wine)

做法

咖啡奶凍：

1. 淡忌廉、鮮奶、咖啡粉及咖啡糖用中火煲滾後離火。
2. 放入魚膠片，攪拌至所有材料溶化，放涼後加入咖啡酒拌勻。
3. 分別注入玻璃杯中，約¼杯份量，放冰箱至凝固。

雲喱拿奶凍：

1. 淡忌廉、鮮奶、砂糖及切開的雲喱拿豆用中火煲滾後離火，加蓋待15分鐘，以便雲喱拿豆出味。
2. 放入魚膠片攪拌至溶化，過篩，以隔去雲喱拿豆，放涼後加入杏仁酒拌勻。

組合：

1. 把雲喱拿奶凍注入已凝固的咖啡奶凍，放冰箱凝固，再倒入咖啡奶凍材料，雪至凝固，重複以上步驟，放冰箱至凝固。
2. 吃前在奶凍面，放上焦糖脆脆可作裝飾，別具風味。

Method

Coffee Panna Cotta:

1. Put whipped cream, milk, coffee powder and coffee sugar into a pot and bring to boil by medium heat. Remove from heat.
2. Add in gelatine leaf and fold in all ingredients until dissolved. Cool and pour Kahlua in it. Mix to combine.
3. Pour into glasses about ¼ full and chill in the fridge.

Vanilla Panna Cotta:

1. Put whipped cream, milk, sugar and vanilla beans in a pot and bring to boil by medium heat. Cover by lid for 15 min to let the vanilla aroma out.
2. Put gelatine leaf in the milk and stir to dissolved. Sift and discard the beans. Cool. Fold in Disaronno Amaretto and mix together.

Assembly:

1. Pour vanilla panna cotta over solidified coffee panna cotta, chill in the fridge. Again pour coffee panna cotta and chill until solidified. Repeat the above steps and chill in the fridge until all set.
2. Serve with crispy caramel before serve.

Cook's Tips 技術指導

焦糖脆脆做法，請參照焦糖脆脆天使蛋糕（第10頁）。

For the making method of the caramel, please follow the instruction in crispy Angel Cake (refer to p.10).

泰式椰汁西米糕
Thai Coconut Sago Pudding

■ 製作時間：**80分鐘**
■ 份量：**12件**
■ Production Time: **80min**
■ Yield: **12 pcs**

材料

斑蘭葉12片
西米80克
滾水1公升
砂糖40克
馬蹄粒3湯匙
粟粉20克
清水50毫升
椰漿200毫升
砂糖30克
⅛茶匙鹽
12粒粟米粒

Ingredients

12 pieces of pandan leaf
80g sago
1 liter of boiling water
40g sugar
3 tbsp water chestnut powder
20g corn starch
50 ml water
200 ml coconut milk
30g sugar
⅛ tsp salt
12 pieces of corn

做法

1. 斑蘭葉沖淨及抹乾，每片葉修剪成20厘米長條，每隔4厘米剪一刀，但不用剪斷，共分成5格。覆摺成盒型，用書釘夾好。

2. 西米沖淨，放滾水內攪拌至水翻滾，蓋好煮5分鐘，熄火，蓋好再焗10分鐘至透身，倒入篩內沖冷水至透明狀，隔乾水。

3. 把西米放回煲內，加入糖及馬蹄粒，以中火煮至稠身，趁熱把1茶匙西米填入斑蘭葉盒內至半滿，鋪平。

4. 粟粉加水、椰漿、糖及鹽拌勻，放小煲內煮至稠身，煮時要不時攪動，把粟粉糊即時鋪西米面，以粟米粒放面裝飾，放冰箱內雪凍後便可享用。

Method

1. Rinse pandan leaves and wipe well. Trim the leaves into 20 cm strips. Divide each leaf into 5 portions and cut at 4 cm intervals (do not cut off to the end). Then fold into a case and stapled.

2. Rinse sago and put into a pot of boiling water until boiled. Continue to cook with lid for 5 min, turn off the heat and keep covered for further 10 min. When sago becomes crystalline, scald them in cold water and drain.

3. Put sago back in a pot with adding in of sugar and water chestnut cubes and boil by medium heat until thickened. Scoop 1 tsp of sago and half fill each pandan leaf case. Spread flat.

4. Mix water with corn starch and stir with coconut milk, sugar and salt. Put into a pot and bring to boil until thickened. Stir occasionally. Put the corn batter pour over sago and chill in the fridge. Serve.

← 20 cm →

Cook's Tips 技術指導

斑蘭葉在泰式雜貨店有售。

Pandan leaf is available in Thai grocery shops.

椰汁馬豆糕
Coconut and Yellow **Split Pea Pudding**

材料

馬豆80克
清水1公升
大菜10克
滾水750毫升
砂糖250克
粟粉80克
清水120毫升
椰汁160毫升
淡奶60毫升

Ingredients

80g yellow split peas
1 liter of water
10g agar agar
750g boiling water
250g sugar
80g corn starch
120 ml water
160 ml coconut milk
60 ml evaporated milk

■ 製作時間：**40分鐘**
■ 份量：**16件**
■ Production Time: **40 min**
■ Yield: **16 pcs**

✎ Lilian 貼心小語

近年有食肆將馬豆糕改良得非常軟滑，其實這款糕點必定要凍食，因為那是由魚膠粉做的椰汁凍糕變化而來，只要加入少許馬豆或紅豆便成凍豆糕了。但如果喜歡吃豆香的話，傳統用粟粉做的糕點才行，所以齋舖出售的紅豆糕、馬豆糕口感雖是結實，仍很受歡迎，例如旺角八寶齋廚的素食糕點做得不錯，全因那些豆焗得夠鬆軟，夠香綿。

Some restaurants have modified this dessert to become very fine. Actually, this dessert is to be served cold and the making method comes from cold coconut cake made up of gelatine powder with adding of coconut of peas or red beans. It you like the pea flavour, those made up of corn starch corn can satisfy you. I have tried this dessert in a vegetarian restaurant located in Mongkok, the finish of yellow split pea is so soft.

做法

1. 馬豆沖淨，用1公升水以中慢火煮約20分鐘，熄火蓋好再焗10分鐘及至豆鬆軟，隔水備用。

2. 大菜剪碎後浸軟，放750毫升滾水內煮約5分鐘至完全溶化，加入糖再煮溶，熄火。

3. 粟粉用120毫升清水拌勻，徐徐拌入糖水中，開火，不時攪拌煮至濃稠，最後加入熟馬豆、椰汁及淡奶以慢火再煮2分鐘。

4. 把材料倒入已灑過水之盛器中，待涼，放冰箱內冷藏至凝固，取出切件，便成香滑椰汁馬豆糕。

Method

1. Rinse yellow split pea and put in a pot with a liter of water, boil by medium heat for 20 min. Turn off the heat and cover by lid by 10 min until the peas soft. Drain well and set aside.

2. Cut agar agar into pieces and soak in water. Put into a pot of 750 ml water and bring to boil for 5 min until dissolved. Add in sugar and boil until dissolved. Turn off the heat.

3. Dissolve corn starch in 120 ml water, fold in sugar solution. Pour the mixture in a pot and turn on the heat for boiling until thickened. Stir occasionally. Add in cooked yellow split peas, coconut milk and evaporated milk and cook by medium heat for 2 min.

4. Pour all ingredients in a container sprinkled with water and cool. Chill in the fridge until hardened. Remove and cut into pieces. Serve.

宮廷桂花糕
Royal Osmanthus **Pudding**

■ 製作時間：**40分鐘**
■ 份量：**12件**
■ Production Time: **40 min**
■ Yield: **12 pcs**

材料

已浸雪耳碎40克
滾水750毫升
杞子20粒
冰糖120克
糖桂花1湯匙
乾桂花1茶匙
魚膠粉30克
冰水4湯匙

Ingredients

40g chopped snow fungus, soaked
750 ml boiling water
20 pcs medlar seed
120g rock sugar
1 tbsp candied osmanthus
1 tsp dried osmanthus
30g gelatine powder
4 tbsp cold water

做法

1. 雪耳浸透，去蒂，切碎，放半煲滾水內先煮10分鐘，隔水備用。

2. 魚膠粉加冰水拌勻；杞子用熱水略浸，隔水備用。

3. 煲內750毫升滾水加雪耳碎及冰糖以慢火煮溶，加入魚膠粉水拌至完全溶解，再加杞子、糖桂花及乾桂花拌勻。

4. 混合物凍透後便可注入已灑過水盛器中，然後再坐冰水，徐徐攪拌至杞子等物浮面，即半凝固狀態，放入冰箱冷藏至完全凝固，便可切件享用。

Method

1. Soak snow fungus, cut stalk and chop then put into a pot and boil with half pot of water for 10 min. Drain well and set aside.

2. Mix ice water with gelatine powder until even. Soak medlar seeds in hot water for a while and drain well. Set aside.

3. Bring 750 ml of water to boil. Add in snow fungus and rock sugar, cook by medium heat until sugar is dissolved. Again add in gelatine liquid until dissolved, then add in medlar seed, candied osmanthus and dried osmanthus, and mix together.

4. Cool the mixture and pour into a container sprinkled with water. Then place the container over ice water and lightly stir those medlar seeds and osmanthus up to surface. The mixture becomes semi-solidified. Chill it in the fridge until set. Cut into pieces and serve.

◥ Lilian 貼心小語

糖桂花是乾桂花用糖醃製而成，它的清甜花香，使清淡的甜品表現突出，正如上海菜桂花丸子或桂花糖蓮藕等，同樣是材料簡單，如果沒有糖桂花，相信沒有這麼吸引。近年，杞子或桂圓桂花糕很受歡迎，那個不過是用魚膠粉做的凍糕，正因有糖桂花的配合，便美名成宮廷桂花糕了，多麼貴氣。

Osmanthus sugar is the mixture of dried osmanthus and sugar. It has distinctive sweet floral fragrance and can bring the light taste of desserts, such as both Shanghai desserts, Gui Hua Rice Balls and Gui Hua Candied Lotus. Without this dried flower sugar, both desserts seems not attractive and hence this Gui Hua Cake which is a common cold cake made up of gelatine powder. Accompany with osmanthus sugar, this dessert becomes so dainty.

燕窩杞子椰汁糕

Coconut Pudding with Swallow's Nest and Medlar Seeds

材料

杞子2湯匙
即食燕窩3湯匙
白殼雞蛋白2個
魚膠粉40克
砂糖130克
滾水300毫升
凍椰漿120毫升
凍牛奶250毫升

Ingredients

2 tbsp medlar seeds
3 tbsp instant swallow's nest
2 egg white (white shell)
40g gelatine powder
130g sugar
300 ml boiling water
120ml cold coconut milk
250 cold milk

■ 製作時間：**30分鐘**
■ 份量：**16件**
■ Production Time: **30 min**
■ Yield: **16 pcs**

⊠ lilian 貼心小語

這款椰汁凍糕曾經是很受歡迎的甜品，那海綿的質感令人回味，可能因保存和製作有點難度，現在的凍糕多已不用蛋白，雖然軟滑，但少了海綿的感覺，若加入了燕窩及杞子，在母親節做給至愛媽媽吃，倒是不錯。

This coconut cold cake has been very hit for its spongy texture. However, it can not be kept for long and the making method is quite complicated. Hence, no egg white is added in this cold cake. Although the texture is still smooth, there is no spongy effect. You can follow my receipe with swallow's nest and medlar seeds and treat your mothers in Mother's Day.

做法

1. 將椰漿牛奶拌勻，入冰箱冷凍；杞子沖熱水後瀝乾備用。
2. 蛋白放金屬盆中打起至企身，備用。
3. 將魚膠粉與砂糖入袋搖勻，逐少灑下滾水中拌溶，把大滾起之魚膠水大力撞入蛋白中，繼續用蛋拂攪拌，坐冰水中拌至涼凍，再拌入凍椰漿奶至半凝固，最後加入燕窩及杞子拌勻。
4. 把混合物倒入一已灑水盛器中，入冰箱冷藏至完全凝固，切件享用。

Method

1. Mix coconut milk and milk and chill in fridge. Rinse medles seeds with hot water and set aside.
2. Whisk egg white in a metal bowl until stiff peak. Set aside.
3. Shake gelatine powder and sugar in a bag, and sprinkle into boiling water bit by bit. Pour boiled gelatine liquid into egg white quickly and keep to stir by a whisk. Set bowl of egg white mixture over ice water until cool. Then fold in cold coconut milk until semi-solidified and mix in swallow's nest and medles seeds until even.
4. Pour the mixture into a container sprinkled with water and chill it in a fridge until set. Cut in pieces and serve.

Cook's Tips 技術指導

緊記一次過撞入蛋白中，然後繼續拌勻，那麼蛋白便不會因太輕而分層浮面了。

Remember to pour the gelatine liquid into egg white by one go and keep stirring. This help mix two ingredients, otherwise, egg white will be on top as too light.

水晶雜豆凍糕
Crystalline Assorted **Beans Jelly**

■ 製作時間：**90分鐘**
■ 份量：**16件**
■ Production Time: **90 min**
■ Yield: **16 pcs**

材料

三角豆80克

紅腰豆80克

眉豆100克

滾水適量

砂糖4½湯匙

魚膠粉40克

砂糖200克

滾水1公升

Ingredients

80g chick peas

80g kidney peas

100g black-eyed beans

Some boiling water

4½ tbsp sugar

40g gelatine

200g sugar

1 liter of boiling water

⊿ Lilian 貼心小語

這個雜豆凍糕，我在美心酒樓一見到時，就被它的藝術造型吸引了。一件食物仿如一件晶瑩剔透水晶般，加上豆豆屬高纖及高蛋白質，美味又健康，值得推薦給注重健康的人。

The first time I encountered this assorted beans cold jelly, I was attracted by its artistic mood and its crystalline form. Besides, it is a highly recommended dessert which is rich in fiber and protein, and is a healthy food indeed.

做法

1. 把三角豆、紅腰豆、眉豆沖淨，分別浸水至脹大，隔去水份。

2. 將三種豆分別放入三個碗中，每碗加入適量滾水至蓋過豆面，隔水蒸40-50分鐘至豆腍，取出，隔去豆水，各自加入1½湯匙砂糖調味，再蒸5分鐘至糖溶，取出待涼。

3. 把三色豆隔乾水，拌勻，放入已灑過水之膠方盒內，也可獨立放入小袋中，備用。

4. 砂糖與魚膠粉拌勻，灑下滾水中拌至完全溶解，待涼。

5. 把四分一凍魚膠粉水注入雜豆面，放冰箱內雪至凝固，取出，再將餘下魚膠粉水慢慢注入豆中，小心不可讓豆浮起，直至注滿魚膠粉水，放回冰箱內雪過夜至完全凝固，取出，切成方件，便成晶瑩通透之三色豆凍糕。

Method

1. Rinse chick peas, kidney peas and black-eyed beans and soak in water separately. Drain.

2. Put three kinds of beans in three bowls respectively and pour boiling water up to the level that cover beans. Steam for 40-50 min until beans are tender. Remove and drain. Each mix with 1½ tbsp sugar and re-steam for further 5 min until sugar is dissolved. Remove and cool.

3. Drain beans and mix three kinds of bean, put them in a plastic container which has been pre-rinsed by water. Or simply put the assorted beans in a small bag and set aside.

4. Mix sugar and gelatine powder and spread into boiling water until dissolved. Cool.

5. Pour ¼ cool gelatine liquid into assorted beans and chill it in the fridge until hardened. Remove and pour balance gelatine liquid onto beans. Pay attention and do not let beans to float while pouring. Chill the mixture in the fridge over night until all solidified. Remove and cut into cube. Serve.

Cook's Tips 技術指導

這款凍糕做法非常簡單，只要注意豆的軟腍度，切不可以鬆散，至於甜度調校，自己應要試味來決定。

This is a simple cold dessert and the degree of softness among three kinds of beans differs. For the sweetness, you have to test on your own.

雙色糯米糍

Dual-colour Glutinous Rice Balls

材料

糯米皮
糯米粉230克，粘米粉40克
砂糖80克，椰漿180毫升
清水300毫升

調色
吉士粉2茶匙，斑蘭葉汁¼茶匙

餡料
榴槤肉2件，熟芒果2個
椰絲120克

奶黃餡
蛋黃1個，砂糖40克
吉士粉20克，粟粉10克
魚膠粉2茶匙，花奶120毫升
水170毫升，煉奶2湯匙

Ingredients

Glutinous rice wrap
230g glutinous rice powder
40g rice flour, 80g sugar
180 ml coconut milk
300 ml water

Colouring
2 tsp custard powder
¼ tsp pandan leaf juice

Filling
2 piece of durian flesh
2 ripe mangoes
120g shredded coconut

Custard Filling
1 egg yolk, 40g sugar
20g custard powder
10g corn starch, 2 tsp gelatine powder
120 ml evaporated milk
170 ml water, 2 tbsp condensed milk

■ 製作時間：**80分鐘**
■ 份量：**36件**
■ Production Time: **80 min**
■ Yield: **36 pcs**

◥ Lilian 貼心小語

芒果及榴槤都是季節性水果，當時令到又平又靚，最好是做甜品，除了布甸，糯米糍也很受歡迎。我曾試過滿記的芒果糯米糍，很美味，原來另一款綠色的是榴槤味，味道也不錯，其實做法差不多，所以順道介紹給榴槤粉絲，各適其式，不妨試做。

Mango and durian are seasonal fruits. When it is season, the price will be cheap and the quality is good, and it is suitable to make dessert like popular pudding and rice balls. I have tried mango rice ball from Honeymoon Dessert, the taste is very good and so as the durian one. The making method is similar and that is why I introduce durian rice ball to. I hope durian lovers will like it.

做法

1. 將榴槤及芒果分別切出18粒果肉，每粒沾上椰絲，放入冰格備用。

2. **糯米皮**：將材料放大碗內攪拌至幼滑，過篩，把半份粉漿加入吉士粉調色，倒入已掃油及墊上牛油紙長盆中，隔水大火蒸18分鐘，取出，即時灑上少許椰絲，脫離蒸盆；將另半份粉漿加斑蘭葉汁調色蒸熟，便成榴槤皮。

3. **奶黃餡**：蛋黃加糖放碗內攪拌至淡黃色，加入其餘材料攪拌成稀漿，放煲內煮至濃稠，期間不停攪拌，最後拌入煉奶。

4. **組合**：把黃色糯米皮用膠刀切出18份，穿上保鮮手套，把每塊糯米皮鋪上少許奶黃餡及一粒芒果肉，包成圓球形，用手指緊捏收口，去除多餘皮料，搓圓，滾上椰絲。綠色皮包入榴槤餡，完成後入冰箱冷凍便可享用。

Method

1. Cut durian flesh and mangoes into 18 pieces respectively and each coat with shredded coconut. Chill in the freezer.

2. **Glutinous rice wrap:** put ingredients in a bowl, stir until smooth, and sift through a sieve. Put custard powder into half of the batter for mixing colour and pour into a tray lined with baking paper. Steam the batter by high heat for 18 min. Remove from heat and sprinkle with some shredded coconut. Remove from tray. Mix balance half batter with pandan leaf juice for colouring, then steam and durian wrap is formed.

3. **Custard Filling :** add egg yolk and sugar in a bowl and whisk until light yellow. Add in balance ingredients and stir into batter, put batter in a pot and cook until thinkened. Stir occasionally. Again fold in condensed milk.

4. **Assembly :** cut yellow glutinous rice wrap into 18 portions by a plastic knife. Put on clear gloves and stuff each wrap with some custard filling and one piece of mango flesh. Knead in ball shape and seal. Trim excess wrap and knead as a ball, and coat with shredded coconut. For green pandan wrap, stuff with durian filling. Chill it in fridge and serve.

鮮雜果黑珍珠

Fresh Fruits with **Black Boba Balls**

材料

黑珍珠3湯匙
溫水適量
砂糖150克
滾水60毫升
椰漿（伴食）適量
鮮雜果如西瓜、哈蜜瓜、士多
　啤梨、芒果、奇異果等。

Ingredients

3 tbsp boba balls
Some warm water
150g sugar
60 ml boiling water
Some coconut milk
Fresh fruits like water melon,
　Hami melon, strawberry,
　mango and kiwi etc.

- 製作時間：**30分鐘**
- 份量：**6人**
- Production Time: **30 min**
- Serving: **6**

做法

1. 黑珍珠用大碗溫水浸15分鐘至漲大，瀝乾，雪凍。
2. 糖加滾水煮約10分鐘成稀糖漿，待涼。
3. 預備鮮果去皮、切粗粒，雪凍。
4. 食時，凍黑珍珠加糖漿及雜果拌勻，也可淋少許椰漿伴食，增添美味。

Method

1. Soak boba balls in a bowl of warm water for 15 min until expanded, drain well and chill.
2. Add sugar in a pot of boiling water and cook for 10 min until turning to thin syrup. Cool.
3. Peel skin of fruits and chop in large cubes, then chill.
4. Add in black boba balls, syrup and mixed fruits when serve or add in some coconut milk to enhance the flavour.

Lilian 貼心小語

黑珍珠是泰國食品，多用作甜品伴食，聽説是金不換的種子，有潤腸功效，可以多吃；黑珍珠外表似細粒黑芝麻，加水浸過後會漲大，變成外層透明，裏面一點點黑色的，咀嚼感覺是爽口的，最好與鮮果伴食，消暑清熱。而甜品屋松記則以黑珍珠加紅豆沙熱食，像是拌入西米般，非常清潤。

Black boba ball is a kind of Thai food and usually goes with dessert. It is said that it is the seed of sweet basil which has nourishing effect for intestine. Boba ball looks like black sesame, it swell after soaking in water and turns translucent with a black dot inisde. The texture is crunchy and it is nice to go with fresh fruits especially during the summer time, having cooling effect. I have tried this dessert in a local sweet soup shop, Chun Kee, they add boba balls into red bean sweet soup and the texture of boba balls seems like sago, with nourishing effect.

焦糖 / 脆糖粒粒

Crispy **Caramel**

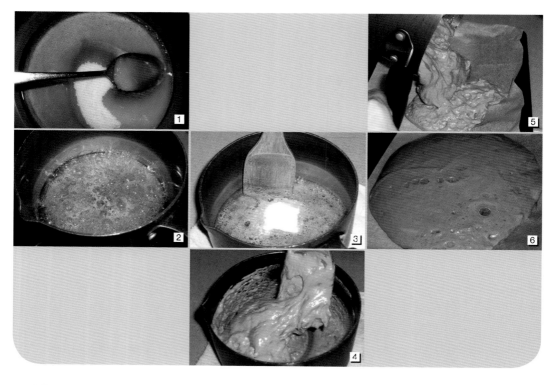

1. 砂糖、清水和蜜芽糖同放進平底鑊中，加熱煮溶。
   Put sugar, water and malt sugar in a pot and boil to dissolve.

2. 糖漿煮至濃郁兼不斷翻滾，出現泡沫。
   Keep boiling the sugar liquid to thickened and frothed.

3. 濃糖漿煮成焦糖色，離火，加入發粉拌勻。
   When thick syrup turns to tea colour, turn off the heat, add in baking soda and stir to combine with a wooden spoon.

4. 焦糖立即產生化學變化，質感變成輕身浮軟。
   The caramel changes to fluffy and light.

5. 倒進已墊牛油紙的焗盤上。
   Pour the caramel mixture on a baking tray lined with baking paper.

6. 待涼凍，貯放在密封盒內，保持乾燥。
   Once the caramel is cool and keep dry into a airtight box.

煮藍莓果醬
Blueberries Jam

1. 藍莓、砂糖、檸檬汁和清水同置煲中烹煮。
   Put blueberries, sugar, lemon juice and water in a pot and bring to boil.

2. 果醬煮至翻騰呈泡沫狀。
   Cook the blueberries purée until frothed.

3. 藍莓仍呈顆粒粒。
   The blueberries remain the form.

4. 當果醬變濃稠時，顆粒仍在，果汁濃郁。
   When the blueberries paste turns thickened, the juice is rich with keeping the berries flesh.

5. 倒入隔篩分離果肉和果汁。
   Separate the flesh and juice by a fine sleve.

6. 用膠刮把藍莓擠壓成茸。
   Use a spatula to crush the blueberries in jam.

7. 果茸經過篩後，變成幼滑狀況。
   After sifiting the blueberry jam, the texture turns to smooth.

蛋黃蓮蓉大壽包
Peach Bun with **Lotus Seed Paste**

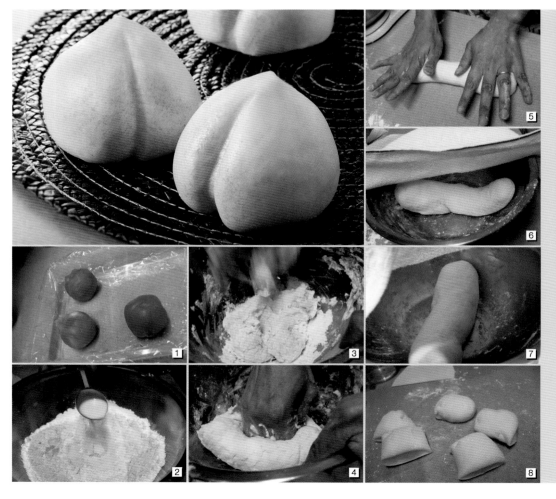

① 蓮蓉包上鹹蛋黃，搓成小圓球。
Stuff salted duck egg yolk into lotus paste and knead into ball.

② 把材料混合。
Mixed up all the ingredients.

③ 用手搓揉材料成糰。
Knead into a dough.

④ 用帶點柔力的拳頭壓搓麵糰。
Use the fist to knead the dough, the effect will be more nice.

⑤ 把麵糰搓長。
Knead the dough into strip.

⑥ 放在盆子內發酵。
Place in a tray and prove.

⑦ 發酵完成，麵糰變大了。
After proving, the dough swells.

⑧ 分割若干份。每個粉糰量重。
Divide the dough into several portions. Weigh each small dough.

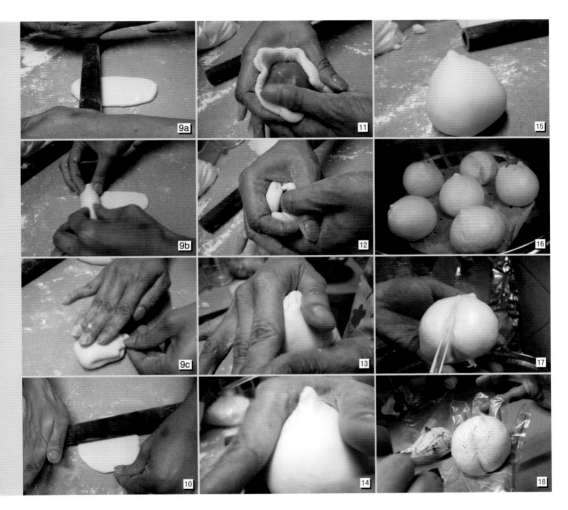

9a-c 把小麵糰碾長，覆摺，再碾長，覆摺。
Roll the dough, fold, roll again and fold.

10 把麵糰碾擀成長圓麵皮。
Knead the dough into thin, oval shape.

11 用麵皮包上蓮蓉餡。
Stuff with lotus seed paste.

12 用手的虎口收捏麵糰。
Seal the end with the part between thumb and index fingers.

13 撕掉多餘麵皮。
Trim excess rim.

14 把頂部弄尖。
Make the top into pointed shape.

15 桃尖向上。
The pointed part faces upward.

16 包子蒸至半熟。
Steam the buns half-cooked.

17 取出包子，用筷子在包身壓出桃坑紋。
Take out the buns and use a chopstick to make a dent, foming peach shape.

18 刷掃蘸點色素，彈在包子上。
Dip some colouring and sprinkle on the buns.

英式楓糖鬆餅
Maple **Syrup Scones**

1 麵糰搓揉後，用保鮮紙包好置冰箱貯存。
Knead the dough, then wrapped by cling firm and chill in the fridge.

2 把麵糰切割，對切成等邊三角形。
Cut the dough into equilateral triangle.

3 將麵糰放在已墊焗餅紙的焗盤上。
Place the dough onto a tray lined with baking paper.

4 掃上楓葉糖漿。
Brush with maple syrup.

5 待烘焙完成時，製品會膨脹和表面轉色。
After baked, the scones will swell and turn into different colour.

6 製品表面光燥而呈光澤。
The top turns dry but with shine.

木糠布甸
Serradura

1. Oreo 餅碎弄好，放在煲子。
   Crush Oreo and place it in a mould.

2. 加入可可粉拌勻。
   Mix in cocoa powder.

3. 布甸弄好，舀在器皿內。
   Make the pudding and place in the mould.

4. 布甸放在器皿，抹平或輕敲器皿至表面平坦。
   Spread the pudding top flat or lightly hit the mould against the table to make the surface flat.

5. 最頂一層為布甸，可放溫室翠苗裝飾。
   Top layer is pudding and can decorate with some greens.

6. 最頂一層是 Oreo 餅碎，也可放溫室翠苗裝飾。
   Top layer is Oreo and can decorate with some greens.

冰花蛋球

Chinese Styled **Donuts**

1. 材料篩勻放在盆中。
   Place all ingredients into a bowl.

2. 加入滾水中，快速攪拌成糰。
   Pour boiling water into the bowl and stir mix.

3. 把麵糰攪至幼滑。
   Mix the dough into smooth condition.

4. 麵糰質感糯軟稀身。
   Texture of the dough turns softened and light.

5. 加入雞蛋後麵糰變得濕潤鬆散。
   The mixture turns into a thick batter after mixing.

6. 麵糰經拂打後變成稀稠的麵糊。
   The mixture turns into a thick batter after mixing.

7. 用雙匙把麵糊弄成球狀，放入熱油炸。
   Use two spoons to scoop balls and deep fry in oil.

8. 蛋球在鑊中受熱膨脹。
   Balls swells in the wok.

9. 蛋球表面出現裂紋。
   Egg buns will have cracks on top.

10. 取出蛋球，放在砂糖內沾糖。
    Remove the egg bun and coat with sugar.

反沙芋條
Candied **Yam**

1. 芋條放入滾油炸熟。
   Deep fry the yam.

2. 取出芋條，瀝油。
   Remove the yam and drain excess oil.

3. 糖水不斷加熱煮沸。
   Boil the sugar solution.

4. 糖水煮至稠身濃郁。
   Continus to boil the sugar solution until frothy and thick.

5. 倒入已炸芋條於糖水中，不斷翻炒。
   Put yam into the sugar solution and keep stir-frying.

6. 芋條滿身沾上砂糖。
   Stop stir-frying when yam coated with sugar.

香港人氣甜品 Most Popular Desserts in Hong Kong

編著　Author
鄭慧芳　胡玉玲　Lilian Cheng　Mindy Wu

編輯　Editor
郭麗眉　謝妙華　Cecilia Kwok　Pheona Tse

翻譯　Translator
布美玲　Jennie Po

攝影　Photographer
幸浩生　Johnny Han

封面設計　Cover Design
羅穎思　Venus Lo

設計　Designer
任霜兒　陸永波　Annie F　BoZ

出版者　Publisher
萬里機構・飲食天地出版社　Food Paradise Publishing Co., an imprint of Wan Li Book Co Ltd.
香港鰂魚涌英皇道1065號東達中心1305室　Room 1305, Eastern Centre, 1065 King's Road, Quarry Bay, Hong Kong.
電話　Tel: 2564 7511
傳真　Fax: 2565 5539
網址　Web Site: http://www.wanlibk.com

萬里機構　萬里 Facebook

發行者　Distributor
香港聯合書刊物流有限公司　SUP Publishing Logistics (HK) Ltd.
香港新界大埔汀麗路36號中華商務印刷大廈3字樓　3/F, C & C Building, 36 Ting Lai Road, Tai Po, N.T., Hong Kong.
電話　Tel: 2150 2100
傳真　Fax: 2407 3062
電郵　E-mail: info@suplogistics.com.hk

承印者　Printer
萬里印刷有限公司　Prosperous Printing Co Ltd

出版日期　Publishing Date
二〇一八年十一月第一次印刷　First Print in November 2018

ISBN　978-962-14-6911-3
Published in Hong Kong